Advances in Aquatic Ecology

Advances in Aquatic Ecology

— Volume 1 —

Editor
Dr. Vishwas B. Sakhare
Post Graduate Department of Zoology
Yogeshwari Mahavidyalaya,
Ambajogai – 431 517
Maharashtra

2007
DAYA PUBLISHING HOUSE
Delhi - 110 035

ISBN 81-7035-483-8

Published by : **Daya Publishing House**
1123/74, Deva Ram Park
Tri Nagar, Delhi - 110 035
Phone: 27383999
Fax: (011) 23244987
e-mail : dayabooks@vsnl.com
website : www.dayabooks.com

Showroom : 4760-61/23, Ansari Road, Darya Ganj,
New Delhi - 110 002
Phone: 23245578, 23244987

Laser Typesetting : **Classic Computer Services**
Delhi - 110 035

Printed at : **Chawla Offset Printers**
Delhi - 110 052

PRINTED IN INDIA

Dedicated to

My Mother

Late Mrs. Sunanda Balasaheb Sakhare

Preface

Unprecedented population growth and rapid industrialization coupled with intensive agricultural and aquacultural practices have exerted intolerable strain on the aquatic resources. Deterioration of water quality, loss of biodiversity and fast depletion of water resources are the main challenges, which need urgent attention. Thus, there is a need of scientific exploitation and conservation of limited aquatic resources. To achieve this goal there is an urgent need of basic and applied research on various aspects of aquatic ecosystems.

The present book is compendium of recent original research in the field of aquatic ecology. It is an assemblage of up-to-date information of rapid advances and developments taking place in the field of aquatic ecology.

The book is a unique compilation of 24 chapters which discusses exhaustive studies on planktonology, physico-chemical environment, fisheries, toxicology, physiology etc. with its application oriented and interdisciplinary approach, the book would be immensely useful to students, teachers, researchers, scientists, policy makers, environmental lawyers and others interested in aquatic ecology.

The chapters in the book have been contributed by eminent scientists/academicians active in the areas of aquatic ecology.

I am highly thankful to Dr. Sureshji Khursale, President, Yogeshwari Education Society, Dr.D.H.Thorat, Principal, Yogeshwari Mahavidyalaya and all the staff members of Yogeshwari Mahavidyalaya, Ambajogai for their encouragement.

I want to thank my wife Surekha for taking constant domestic care bearing all inconveniences during the past few months while the book has been taking its final shape. She has also helped me for constantly data feeding, processing and reprocessing on computer. I want to thank my father Shri Balasaheb Sakhare, brother Avinash, sisters Preeti and Minal for their help in many ways.

I express my gratitude to M/S Daya Publishing House, Delhi, who took interest in my venture and book could see the light of the day.

Thanks are also due to all those who have not been mentioned by name but nevertheless have been of invaluable help in their inscrutable ways.

Dr. V.B. Sakhare

Contents

List of Contributors

Ajmal Khan, S.
Center for Advanced Study in Marine Biology, Annamalai University, Parangipettai – 608 502, Tamil Nadu

Bama Devi, B.
Post Graduate Department of Zoology, Manikam Lutheran College, Porayar – 609 307, Tamil Nadu

Benarjee, G.
Fisheries Research Lab, Department of Zoology, Kakatiya University, Warangal – 506 009, Andhra Pradesh

Bharti, Pawan Kumar
Department of Zoology and Environmental Science, Gurukula Kangri University, Hardwar – 249 404, Uttaranchal

Charak, K.S.
Chief Project Officer, Department of Fisheries, Fisheries Marketing Division, Narwal, Jammu – 180 006
E-Mail: kulwantcharak@indiatimes.com

Chari, M.S.
Department of Fisheries, College of Agriculture, Indira Gandhi Agricultural University, Raipur – 492 006, Chhatisgarh

Das, Surjit
Center for Advanced Studies in Marine Biology, Annamalai University, Parangipettai – 608 502, Tamil Nadu
E-Mail: surajit_cas@yahoo.com; surajit@myself.com

Deecaraman, M.
Department of Biotechnology, Dr. M.G.R. Deemed University, Chennai – 600 095, Tamil Nadu

Deivasigamani, B.
CAS in Marine Biology, Annamalai University, Parangipettai – 608 502 , Tamil Nadu
Email: sigamani_bd@rediffmail.com

Fayaz, F.A.
Project Officer, Department of Fisheries, Regional Fish Farmers Development Agency, Ghomanahan, Jammu

Gaur, Sandhya Rani
Department of Fisheries, College of Agriculture, Indira Gandhi Agricultural University, Raipur – 492 006, Chhatisgarh

Ghosh, Indranil
West Bengal University of Animal and Fishery Sciences, Kolkata – 700114, West Bengal
E-Mail: ig_1970@yahoo.co.in; ind@indiainfo.com

Gopal Samy, A.
Tamil Nadu Organic Farmers Movement (TOFARM), Nagapattinam, Tamil Nadu
E-Mail:jgnagai@yahoo.co.in

Harish Kumar, K.
Department of Environmental Science, Bioscience Complex, Kuvempu University, Shankaraghatta – 577 451, Karnataka

Jayakumar, S.
Department of Zoology, Pachaiyappa's College, Chennai – 600 030, Tamil Nadu

Karpagam, S.
Department of Botany, Queen Mary's College, Chennai – 600 004, Tamil Nadu

Ketkar, Sampada
B404, Balwant Nivas, Sadashiv Peth, Pune – 411 030, Maharashtra
E-mail: sampada.v.ketkar@gmail.com

Khabade, S.A.
Department of Zoology, D.K.A.S.C. College, Ichalkaranji, District Kolhapur, Maharashtra

Kiran, B.R.
Department of Environmental Science, Bioscience Complex, Kuvempu University, Shankaraghatta – 577 451, Karnataka

Kulkarni, K.M.
Director, Higher Education, Pune, Maharashtra

Kumar, Adarsh
Soil and Water Analysis Laboratory, Department of Applied Aquaculture, Barkatullah University, Bhopal – 26, Madhya Pradesh

Laxma Reddy, B.
Fisheries Research Lab, Department of Zoology, Kakatiya University, Warangal – 506 009, Andhra Pradesh

Lyla, P.S.
Center for Advanced Studies in Marine Biology, Annamalai University, Parangipettai – 608 502, Tamil Nadu

Mahajan, S.R.
Department of Botany, Bhusawal Arts, Science and P.O.N. Commerce College, Bhusawal – 425 201, Maharashtra

Majhi, Sullip K.
ICAR Research Complex for NEH Region, Umroi Road, Barapani – 793 103, Meghalaya
E-mail: sullipkm@gmail.com

Malik, Devendra S.
Department of Botany and Microbiology, Gurukula Kangri University, Hardwar – 249 404, Uttaranchal

Mastan, S.A.
Soil and Water Analysis Laboratory. Department of Applied Aquaculture, Barkatullah University, Bhopal – 26, Madhya Pradesh

Mule, M.B.
Department of Zoology, Shivaji University, Kolhapur, Maharashtra

Nandan, S.N.
PG Department of Botany, S.S.V.P.S's L.K. Dr. P.R. Ghogrey Science College, Dhule – 424 005, Maharashtra

Pejaver, Madhuri
B.N. Bandodkar College of Science, Chendani, Koliwada, Thane, Maharashtra

Pratima, M.
Department of PG Studies and Research in Botany, Gulbarga University, Gulbarga – 58 5104, Karnataka

Puttaiah, E.T.
Department of Environmental Science, Bioscience Complex, Kuvempu University, Shankaraghatta – 577 451, Karnataka

Qureshi, T.A.
Soil and Water Analysis Laboratory, Department of Applied Aquaculture, Barkatullah University, Bhopal – 26, Madhya Pradesh

Raja, I.A.
Department of Zoology, Shri Shivaji College of Arts, Commerce and Science, Akola – 444 001, Maharashtra
E-Mail: rasbora _raja @ rediffmail.com

Rajasegar, M.
Post Graduate Department of Zoology, Tranquebar Bishop Manikam Lutheran College, Porayar – 609 307, Tamil Nadu
E-Mail: drravinila@rediffmail.com

Salgotra, Vikas
Soil and Water Analysis Laboratory, Department of Applied Aquaculture, Barkatullah University, Bhopal – 26, Madhya Pradesh

Salim, Sultan
Managing Director, Fishermen's Co-op Federation, Nirala Nagar, Lucknow, Uttar Pradesh
E-Mail: up.fish@up.nic.in

Sharma, Dushyant Kumar
Department of Zoology, Govt. P.G. College, Morena, Madhya Pradesh

Shiddamallayya, N.
Regional Research Institute (Ay.), Ashoka Pillar, Jayanagar, Bangalore – 560 011, Karnataka
E-Mail: siddamallayya_matapati@rediffmail.com

Singh, Deepak Kumar
Department of Botany and Microbiology, Gurukula Kangri University, Hardwar – 249 404, Uttaranchal

Singh, Pradeep Kumar
Department of Fisheries, College of Agriculture, Indira Gandhi Agricultural University, Raipur – 492 086, Chhatisgarh
E-Mail: singhpk_aqua@yahoo.co.in; singhpk_aqua123@rediffmail.com

Srinivasan, M.
Center for Advanced Study in Marine Biology, Annamalai University, Parangipettai – 608 502, Tamil Nadu

Suryawanshi, G.D.
Post Graduate and Research Department of Zoology, Yogeshwari Mahavidyalaya, Ambajogai – 431 517, Maharashtra

Venkateswara Rao, A.
Manager–Technical Services, Aquaculture Products Division, Neospark Drugs and Chemicals Pvt. Ltd., Hyderabad – 500 082, Andhra Pradesh

Chapter 1

Ecology of Sahastradhara Hill-Stream at Dehradun, Uttaranchal

D.S. Malik and P.K. Bharti

Introduction

Himalayas are the cradle of a large number of streams and mighty rivers. Lesser Himalayas consists of a main part of Uttaranchal *i.e.* Kumaon and Garhwal Himalaya. Dehradun, Hardwar and Rishikesh are the main religious and beautiful cities, situated in the foothills of Garhwal Himalaya. A hill-stream Sahastradhara, a major tributary of river Song flows downward through Dehradun valley.

Dehradun valley rests in the arms of the Himalaya in the North and Shiwalik in the South and transversely bordered by Ganga in the South-East. Yamuna in the North-West forms an intermontane valley ecosystem extending from 29° 55′ to 30° 30′ N latitude and 77° 35′ to 78° 20′ E longitude with the area of about 2000 square kilometer ranging from 330 meter to 1800 m above msl (Valdiya, 1975).

The main sources of water in Himalayas are glaciers. The glaciers are an important resource of fresh water stored in the solid state and regulated by climate. Glacier melts and the water pass through heavy and porous sediment materials in the form of river. Water is the basic requirement of all living organisms. Lotic habitats are those exiting in relatively fast running streams, springs, river and brooks. The lentic habitats represent lakes, marshes, swamps, ponds and wetlands. A hill-stream consists of two major zones, rhithron and potamon.Aquatic biodiversity is mainly occurring in potamon zone due to the slow velocity of running water.

Ecological study of hill stream describes the distribution and abundance of aquatic living organisms and also their contribution into environment. Biological diversity mainly influenced by the physico-chemical characteristics of the aquatic ecosystem through the food chain. The biotic community in the stream exhibited the quantitative and qualitative fluctuations due to varied nutrient dynamics and ultimately changed the different trophic status of species community and limnological status of aquatic ecosystem.

Physico-chemical and biological parameters play an important role in population dynamics of phytoplankton, zooplankton, benthos and fishes. Various parameters that affect the population dynamics of phytoplankton may be broadly classified as light intensity, photoperiod, chlorophyll content, nutrient availability, transparency, turbidity, temperature, velocity, water column depth. Light intensity, duration and water column depth are the basic parameters and others are the derived ones (Khanna *et al.*, 2005)

The present study deals with biological diversity, primary productivity and physico-chemical characteristics of Sahastradhara stream, which gives a baseline data about the biological status of stream. The purpose of this work is to provide preliminary observations of the ecology of natural water from Sahastradhara at Dehradun and to make geo-chemical comparison between 'Sahastradhara' water and water from other regions of the country. The present work is aimed towards developing baseline data for river water quality of *Sahastradhara* stream, a major tributary of Song River and to understanding the ecological processes active in Himalayan river basins.

Materials and Methods

The sampling site is situated in Sahastradhara valley, 15 km far from Dehradun, selected five sampling stations along with stream for limnological study. The water samples were collected monthly from different sampling site s_1, s_2, s_3, s_4 and s_5 during 2003 to 2004 in morning period (from 9.00 am to 11.00 pm). First upstream S_1 (Karligarh village), second upstream S_2 (Parking bridge), main tourist spot S_3 (Sulphur Spring), first down stream S_4 (Crossing bridge) and second downstream S_5 (Bus stand) were the sampling stations with a distance of about 1 km for each. The samples were taken in BOD bottles and plastic Jeri cane for physical, chemical and biological parameters. The samples for different parameter were analyzed with the help of the procedure described by Mark (1977), APHA (1995), Mathur (1982), Trivedi and Goel (1984), Santhanam *et al.* (1989), Admondson (1992) and Khanna (1993).

Results and Discussion

The impact on water quality of anthropogenic pollution from industrial, agricultural, sources, quarrying and tourists activity has concerned environmentalists and scientists for the past two decades. Plankton is the primary producer of any aquatic ecosystem, which can be as an indicator of water quality. Dissolved oxygen decreases or increases in an aquatic ecosystem due to the plankton. Plankton takes sunlight directly and makes food for the total ecosystem through the photosynthesis process. Plankton may vary at place-to-place and season-to-season.

The mean values of physical and chemical parameters of Sahastradhara stream are depicted in Table 1.1 and 1.2 respectively. Temperature plays a vital role in all physico-biochemical reactions and self-purification power of aquatic system. Organic matter, which is oxidized, is highly influenced by the temperature. Temperature fluctuation measurement is useful to determine the trends of bio-chemical and biological activities. The Sahastradhara stream has a difference in the fluctuation of water temperature was recorded maximum (17.7°C±0.4) in the month of May and minimum (13.1°C±0.66) in January. The temperature of water decreases from September and increases from January onward. Badola and Singh (1981) observed similar trend in the river Alaknanda in the Garhwal region. The maximum value of velocity of Sahastradhara stream was recorded in the month of May (0.64m/s±0.257) and minimum (0.58m/s±0.1315) in December. During the study period it has been observed that from January onward the velocity starts increasing up to April. Similar observation has been recorded by Joshi *et al.* (1995) in western Ganga canal at Hardwar.

Table 1.1: Physical Parameters (Mean Values) of Sahastradhara Stream

Parameters	*Site–I*	*Site–II*	*Site–III*	*Site–IV*	*Site–V*
Temperature (°C)	15.75±1.14	15.37±1.63	15.0±1.41	15.75±1.6	15.75±2.277
(Range)	(14–18)	(13–17.5)	(13–17)	(13.5–18)	(12–18)
Velocity (m/s)	0.3±0.14	0.9±0.1	0.35±0.05	0.8±0.02	0.65±0.05
(Range)	(0.3–0.4)	(0.28–0.32)	(0.8–1.0)	(0.78–0.82)	(0.6–0.7)
Turbidity (JTU)	2.75±1.47	3.25±1.30	4.5±1.15	3.75±1.48	4.75±1.48
(Range)	(1.0–5.0)	(2.0–5.0)	(3.0–6.0)	(2.0–6.0)	(3.0–7.0)
Total solids (mg/l)	1218±183.2	1234.2±89.8	1277.25±122.6	1305.7±166.1	1290.75±146.1
(Range)	(955–1427)	(1090–1332)	(1115–1443)	(1135–1520)	(1148–1535)
TDS (mg/l)	1093.25±175	1094.25±120.5	1140.5±134.6	1189±158.01	1134.25±176.1
(Range)	(840–1278)	(895–1217)	(1010–1322)	(1039–1380)	(965–1420)
TSS (mg/l)	124.75±14.39	140.0±28.16	136.75±40.95	116.75±13.8	156.5±41.69
(Range)	(113–149)	(119–195)	(105–207)	(105–140)	(115–211)

The seasonal fluctuation of pH depends chiefly on amount of free carbon-dioxide, which is ultimately related to the rate of photosynthesis and also the decomposition of organic matter. The pH of natural water was controlled in a greater extent by the interaction of hydroxyl ions, resulting form the dissociation of carbonic acid and form hydroxyl ion arising from the hydrolysis of bicarbonate. In this study, it was recorded that the pH was always slightly alkaline. The pH of stream water was recorded highest (8.02±0.1166) in December and lowest (7.4±0.1414) in the month of April. Various workers in their studies have shown similar results in other rivers of Garhwal region. Sharma (1986) has also indicated the same trend of pH values on seasonal basis in the Bhagirathi River, Garhwal Himalayas.

Table 1.2: Chemical Parameters (Mean Values) of Sahastradhara Stream

Parameters	*Site–I*	*Site–II*	*Site–III*	*Site–IV*	*Site–V*
pH	7.625±0.43	7.6±0.28	7.75±0.15	7.75±1.66	7.6±0.339
(Range)	(7.2–8.1)	(7.3–8.1)	(7.5–7.9)	(7.6–8.0)	(7.2–8.1)
Free CO_2 (mg/l)	1.6525±0.39	1.75±0.017	1.825±0.25	1.74±0.09	1.87±0.21
(Range)	(1.40–1.69)	(1.62–1.92)	(1.62–2.20)	(1.65–1.85)	(1.62–2.20)
D O (mg/l)	8.41±1.62	8.46±1.18	8.65±0.95	7.55±0.87	7.49±1.36
(Range)	(7.04–11.08)	(6.84–10.07)	(7.04–9.47)	(6.04–9.06)	(6.04–9.47)
BOD (mg/l)	1.875±0.162	1.98±0.23	2.25±0.18	2.0±0.14	2.05±0.15
(Range)	(1.60–2.0)	(1.7–2.2)	(2.0–2.5)	(1.8–2.2)	(1.9–2.3)
COD (mg/l)	2.55±0.32	2.6±0.25	2.9±0.22	2.68±0.22	2.7±0.18
(Range)	(2.1–3.0)	(2.2–2.9)	(2.6–3.2)	(2.4–3.0)	(2.4–2.9)
Hardness (mg/l)	196.25±8.2	200.0±15.41	222.5±19.2	200.0±18.71	198.75±21.02
(Range)	(190–210)	(185–220)	(200–250)	(180–230)	(185–235)
Alkalinity (mg/l)	250.0±134.7	262.5±140.8	305±172.1	292.5±168.8	317.5±175.2
(Range)	(150–480)	(140–500)	(170–600)	(130–570)	(120–540)

The dissolved oxygen showed an inverse relationship with temperature as reported by Das and Srivastava (1956) and Dobriyal (1985). The dissolved oxygen in water is often attributed to the fact that the oxygen is dissolved more during the period of active photosynthesis. Hutchinson (1957) reported that when the water has low temperature has greater capacity for holding dissolved gases. The highest value of dissolved oxygen was recorded (9.85 mg/l±0.776) in the month of January and the lowest value of dissolved oxygen was observed (6.68 mg/l±0.387) in the month of May. This trend was also stated by Khanna *et al.* (1992) in Ganga river at Hardwar. The higher value of dissolved oxygen during winter months can be attributed to increased solubility of oxygen at lower water temperature and high photosynthetic rate.

Free carbon dioxide released during the decomposition of certain substances and respiration of living organism since higher temperature accelerate the decomposition of organic substances as well as the respiratory activity of the biota and shown direct relationship between the water temperature and free carbon dioxide (Hynes, 1970). The maximum value of free carbon dioxide was observed (2.2 mg/l±0.155) in May and minimum value in January (1.70 mg/l±0.066). Free carbon dioxide and dissolved oxygen showed a negative relationship to each other. The similar observations have also been reported by Quadri and Shah (1989) in some fresh water ecosystem. The highest rate of photosynthesis decreases free carbon-dioxide content and higher rate of decomposition increases free carbon dioxide. An inverse relationship between dissolved oxygen and free carbon-dioxide has been reported by Khan and Siddiqui (1971).

Chloride was found in form of chloride ion, and one of the major inorganic anions present in natural water. The maximum chloride concentration was recorded (17.14mg/l±1.33) in the month of May and minimum in the month of January (11.36 mg/l±2.008), chloride and hardness showed a positive relationship to one another. Chopra and Patrick (1994) also stated similar trend in aquatic ecosystem in the river Ganga at Rishikesh. The maximum calcium concentration was observed (15.99mg/l±5.53) in the month of May while the minimum value of calcium was observed (80.336mg/l±2.64) in the month of January. Similar type of relationship of calcium was observed by Ahemad and Hasnain (2001) in fresh water bodies. Khanna and Singh (2000) also reported similar trend of calcium distribution in river Suswa, Raiwala, Dehradun. Sahastradhara stream may be regarded as rich in calcium and other geochemical nutrients. Calcium is one of the essential nutrients, which plays an important role in biological system. In fact, calcium content in water is responsible for bone growth of children.

Table 1.3: Nutrients Parameters (Mean Values) of Sahastradhara Stream

Parameters	*Site–I*	*Site–II*	*Site–III*	*Site–IV*	*Site–V*
Chloride (mg/l)	12.78±1.73	13.73±1.93	15.26±2.1	13.49±3.25	11.6±3.06
(Range)	(11.36–15.62)	(11.36–16.58)	(12.78–18.46)	(9.94–18.46)	(8.52–16.58)
Calcium (mg/l)	84.67±3.58	85.67±4.28	91.68±10.13	89.18±8.26	88.43±6.19
(Range)	(80.160–88.176)	(80.160–92.184)	(77.154-104.208)	(79.158-99.198)	(80.16–96.192)
Magnesium (mg/l)	48.66±3.09	57.33±3.29	64.66±2.36	55.33±2.49	51.0±5.29
(Range)	(45–53)	(55–62)	(63.0–68.0)	(52.0–28.0)	(45.0–55.0)
Sodium (mg/l)	7.375±0.96	7.725±0.506	8.75±0.75	8.87±1.14	9.25±1.35
(Range)	(6–8.5)	(7.0–8.2)	(8.0–9.5)	(7.0–10.0)	(7.5–11.0)
Potassium (mg/l)	15.13±0.74	15.75±0.901	16.62±0.65	16.5±0.79	17.0±1.35
(Range)	(14–16)	(14.5–17.0)	(16.0–17.5)	(15.5–17.5)	(16.0–18.0)

Table 1.4: Correlation (=CORREL) Between Nutrients (Mean Values) and Physico-chemical Parameters (Mean Values) of Sahastradhara Stream

Parameters	*Chloride*	*Calcium*	*Magnesium*	*Sodium*	*Potassium*
Temperature (°C)	– 0.86097	– 0.51833	– 0.92056	– 0.01186	– 0.14275
Velocity (m/s)	– 0.17595	– 0.12077	0.007842	0.179873	0.225839
Turbidity (JTU)	0.012697	0.818834	0.400371	0.92284	0.959357
Total solids (mg/l)	– 0.00448	0.788427	0.273292	0.949891	0.906666
TDS (mg/l)	0.133154	0.71132	0.25802	0.764443	0.682478
TSS (mg/l)	– 0.35586	0.096981	0.004271	0.35689	0.46296
pH (pH unit)	0.681329	0.761113	0.623067	0.395679	0.345766
Free CO_2 (mg/l)	– 0.03224	0.686443	0.396104	0.833271	0.90927
DO (mg/l)	0.679888	– 0.12231	0.471533	– 0.65112	– 0.55633
BOD (mg/l)	0.607899	0.912138	0.845587	0.615583	0.694262
COD (mg/l)	0.595229	0.954861	0.801927	0.654354	0.704159
Hardness (mg/l)	0.814816	0.795954	0.897315	0.312648	0.386789
Alkalinity (mg/l)	– 0.04111	0.829565	0.331085	0.97707	0.985243

Magnesium is also an essential and beneficial element, but it is toxic at higher concentration. The recommended acceptable limit of magnesium for domestic purpose is 30mg/l in the river. The amount of magnesium in the river Sahastradhara was found maximum (57.2mg/l±7.24) in the month of May while the minimum value of magnesium was recorded (52.2mg/l±6.55) in the month of January. Castante *et al.* (1985) in river Parana and Meybeck (1984) in river Zaire also reported similar trend of magnesium availability in natural water.

Sodium at lower concentration has no adverse effect on human health. The maximum sodium concentration was recorded (9.44 mg/l±1.016) in the month of May and minimum value of sodium was observed (7.1mg/l±0.663) in the month of January. Pande and Mishra (2000) observed similar results on Sahastradhara stream. Sodium is one of the most common cation present in the water bodies. Potassium is naturally occurring elements, like the sodium, however its concentration remains lower than sodium. The potassium is released by the clay mineral, weathering and leaching form the growing vegetation and decomposition of organic matter. Potassium concentration was observed highest in the month of May (17.2 mg/l±0.67) and lowest in the month of January (15.2 mg/l±0.812). Similar results were obtained by Sarin *et al.* (1989) and Subramanian (1989) observed the similar trend on the river Brahmaputra.

The quantitative fluctuation of the planktonic diversity has shown a significant effect on the ecology of river. All natural water contains a variety of plankton as the natural flora and fauna. The phytoplankton community is a heterogeneous group of tiny plants adapted to suspension in the fresh water. The present study showed that the maximum number of total plankton were recorded (1523.6±26.56) in the month of January and minimum (718.6±19.33) in the month of May.

Phytoplanktons were recorded maximum (1371±25.77) in the month of January and minimum (633±9.32) in the month of May. Eddy (1993) pointed out that plankton production is mainly influenced by temperature. Badola and Singh (1981) reported high values of plankton during January to March in freshwater of Kumaon Lake. According to Das and Pathani (1978) increment in turbidity will reduces the plankton production in a water body due to low light penetration and low rate of photosynthesis. According to Khanna (1993) the plankton were maximum in the month of winter probably due to low temperature, high content of dissolved oxygen low velocity and transparency of water and other suitable conditions. Allen (1920) stated that water currant is the chief factor influencing the plankton of stream San Joaquin River and its tributaries near Stockton in California.

Table 1.5: Planktonic Population (Mean Values) of Sahastradhara Stream

Parameters	*Site–I*	*Site–II*	*Site–III*	*Site–IV*	*Site–V*
Total plankton (n/l)	1130.25±316.8	1133±316	1101.5±323.9	1063.75±346.9	1163.25±339.4
(Range)	(700–1540)	(725–1525)	(693–1480)	(700–1490)	(730–1583)
Phytoplankton (n/l)	1018.25±286.1	1023.75±291.5	992.5±296.3	951.25±318.2	1037.5±259.5
(Range)	(625–1380)	(635– 1380)	(615–1345)	(635–1340)	(640–1410)
Zooplankton (n/l)	112±31.61	109.25±27.09	109±28.6	112.5±31.72	125.75±36.66
(Range)	(75–160)	(77–145)	(78–140)	(65–150)	(90–173)
Total diatoms (n/l)	764.75±206.6	784.5±218.5	772.25±247.3	732.5±236.5	807±250.5
(Range)	(492–1022)	(645–1015)	(480–1064)	(495–1035)	(495–1108)
Green algae (n/l)	190.5±59.86	179.5±64.96	161±40.72	164.5±57.21	157.5±50.13
(Range)	(105–275)	(98–278)	(102–205)	(105–228)	(102–220)
Blue green algae (n/l)	63±24.53	57.25±21.57	59.25±17.72	54.25±25.86	58±32.04
(Range)	(21–83)	(27–87)	(33–76)	(27–83)	(23–97)

Zooplanktons were recorded maximum (152.6±13.00) in the month of December and minimum (82±13.02) in the month of May. Total diatoms were recorded maximum (1048.8±34.02) in the month of January and minimum (495±9.25) in the month of April. Green algae were recorded maximum (241.2±29.77) in the month of January and minimum (106.4±6.28) in the month of May. Blue Green algae was recorded maximum (81±4.05) in the month of January and minimum (27.6±3.98) in the month of May. Juttner *et al.* (1995) also reported the similar trend of diatom communities in Himalayan Hill Streams. Similar observations have been reported by Khanna *et al.* (1998) in relation to phytoplanktonic communities of Ganga canal at Haridwar. Khanna and Singh (2000) also reported the similar trend of planktonic diversity in Suswa River, Raiwala, Dehradun. Similar observations were also reported by Nautiyal (1982) in Alaknanda River and Khanna (1993) in river Ganga at Hardwar.

Table 1.6: Primary Productivity (Mean Values) of Sahastradhara Stream

Parameters	*Site–I*	*Site–II*	*Site–III*	*Site–IV*	*Site–V*
Net primary productivity (mg/l)	0.271±0.069	0.220±0.0215	0.336±.108	0.231±0.086	0.265±0.0522
(Range)	(0.200–0.355)	(0.200-0.255)	(0.140–0.500)	(0.105–0.320)	(0.200–0.330)
Gross primary productivity (mg/l)	0.6925±0.074	0.640±0.0776	0.716±.1344	0.729±0.0975	0.676±0.0826
(Range)	(0.605–0.805)	(0.600–0.705)	(0.550–1.01)	(0.505–0.900)	(0.580–0.810)
Community respiration (mg/l)	0.420±0.0184	0.420±0.0635	0.421±0.0394	0.498±0.156	0.411±0.0375
(Range)	(0.405–0.450)	(0.400–0.450)	(0.390–0.480)	(0.400–0.580)	(0.380–0.480)
Net production efficiency (per cent)	38.77±4.7	34.29±1.101	44.15±15.56	30.485±6.445	38.87±3.069
(Range)	(33.05–44.09)	(33.33–36.17)	(25.45–66.66)	(20.79–37.03)	(34.48–42.55)

It is well established that the primary productivity is controlled by several physico-chemical and biotic interactions. The rate of production of organic matter per unit time is termed as productivity. There are two levels of production, primary and secondary. The amount of new organic matter resulting from photosynthesis is termed as primary productivity. The animal and plant communities inhabiting water bodies are considered as the direct manifestation of the productivity. The maximum net primary productivity was observed (0.335±0.093) in the month of December and minimum (0.170±0.04) in the month of May. The maximum gross primary productivity was observed (.796±.123) in the month of January and minimum (.569±0.0379) in the month of May. The maximum net community respiration was observed (.479±0.0575) in the month of January and minimum (.399±0.0102) in the month of April. The higher productivity in the present studied was concomitant with temperature range from 14°C to 18°C. Singh and Swaroop (1980) observed positive correlation with productivity and water temperature. A positive correlation existed between gross primary productivity and calcium in Cavery (Lakshminarayan, 2001). Vashney *et al.* (1983) present study a positive correlation was recorded between gross primary productivity and dissolved oxygen while the negative correlation between gross primary productivity and temperature and free carbon dioxide.

Table 1.7: Correlation (=CORREL) Between Plankton (Mean Values) and Physico-chemical Parameters (Mean Values) of Sahastradhara Stream

Parameters	*Total Plankton*	*Phytoplankton*	*Zooplankton*	*Total Diatom*	*Green Algae*	*Blue Green Algae*
Temperature (°C)	0.140144	0.040568	0.561048	– 0.12814	0.203944	– 0.06004
Velocity (m/s)	– 0.09075	– 0.11978	0.096464	– 0.00984	– 0.18658	– 0.81972
Turbidity (JTU)	0.103772	0.005161	0.537763	0.394692	– 0.95776	– 0.35417
Total solids (mg/l)	– 0.40534	– 0.52983	0.405479	– 0.18218	– 0.91516	– 0.73149
TDS (mg/l)	– 0.70298	– 0.80425	0.139674	– 0.54607	– 0.71164	– 0.72652
TSS (mg/l)	0.823334	0.77943	0.635537	0.966082	– 0.40829	0.080955
pH	– 0.88664	– 0.89145	– 0.42828	– 0.71618	– 0.42595	– 0.36483
Free CO_2 (mg/l)	0.27888	0.196314	0.548124	0.578174	– 0.89315	– 0.36232
DO (mg/l)	0.028383	0.184226	– 0.75181	0.04589	0.505435	0.574037
BOD (mg/l)	– 0.18081	– 0.18513	– 0.07095	0.161374	– 0.74373	– 0.20002
COD (mg/l)	– 0.27454	– 0.29116	– 0.05821	0.047125	– 0.75928	– 0.19054
Hardness (mg/l)	– 0.2978	– 0.25029	– 0.38539	– 0.01229	– 0.45865	0.023773
Alkalinity (mg/l)	– 0.00781	– 0.12433	0.568959	0.262619	– 0.98348	– 0.44697

Net production efficiency shows the positive correlation with dissolved oxygen, while the negative correlation between net production efficiency and carbon dioxide. The maximum mean value of net production efficiency was observed (43.79 per cent) in the month of January and minimum (29.88 per cent) in the month of April. Similar trend was observed by Yeragi and Shaikh (2003) on the primary productivity of Tansa River. Singh *et al.* (1980) and Kar *et al.* (1978) however, recorded low and high productivity in summer and winter respectively.

Sahastradhara stream has a wide biodiversity of small invertebrates, including some benthos, nekton, phytoplankton, zooplankton, which maintain the water quality of aquatic ecosystem. Planktonic genus and other fauna available in the water of Sahastradhara hill stream at Dehradun depicted in Table 1.9.

Table 1.8: Correlation (=CORREL) Between Primary Productivity (Mean Values) and Physico-chemical Parameters (Mean Values) of Sahastradhara Stream

Parameters	*Net Primary Productivity*	*Gross Primary Productivity*	*Community Respiration*	*Net Production Efficiency*
Temperature (°C)	– 0.59945	0.006189	0.311429	– 0.57232
Velocity (m/s)	– 0.82147	– 0.40554	0.387568	– 0.78591
Turbidity (JTU)	0.459379	0.237629	– 0.09751	0.389256
Total solids (mg/l)	0.086601	0.565255	0.544701	– 0.16675
TDS (mg/l)	0.01787	0.761621	0.801179	– 0.33031
TSS (mg/l)	0.165469	– 0.58282	– 0.73453	0.444083
pH	0.409892	0.87509	0.647625	0.023325
Free CO_2 (mg/l)	0.335796	– 0.03623	– 0.24934	0.360837
DO (mg/l)	0.43596	– 0.20694	– 0.47748	0.522691
BOD (mg/l)	0.715985	0.330189	– 0.11921	0.583037
COD (mg/l)	0.757185	0.484589	– 0.01968	0.572202
Hardness (mg/l)	0.811178	0.38677	– 0.13668	0.659382
Alkalinity (mg/l)	0.380434	0.349649	0.067672	0.259953

Table 1.9: Preliminary Study of Biotic Communities in Sahastradhara Stream, Dehradun

Family/Phylum	*Genus/Species*
Flora	
A) Chorophyceae:	*Ulothrix, Volvox, Cladophora, Chlorella, Vaucheria, Zygnema, Pediasterum*
B) Bacillariophyceae:	*Diatoms*
C) Rhodophyceae:	*Batrachospermun*
D) Cyanophyceae:	*Rivularia, Nostoc*
Fauna	
A) Protozoans:	*Vorticalla, Amoeba, Paramoecium, Chrysamocha, Volvox*
B) Coelentrata:	*Hydra, Ceratella*
C) Annelida:	*Pheretima*
D) Arthropods:	*Mayfly nymph*
E) Macroinvertebrates:	*Snails, Leeches, Flatworms, Beetles*

The endemic fish species about 18 in number occurs in the natural aquatic system of Sahastradhara (Table 1.10). Among these, four to six species may be undertaken as sport fishery to enhance or attract the fish anglers on the tourist sites. The limnological and biological status of Sahastradhara stream have shown optimum suitable environment to accelerate the growth and population of endemic fish diversity. There has taken place to increasing use of aquatic resources to cater the needs of tourism. As the present conditions are in sites development of these water bodies has to be taken up for the protection of natural habitats and rehabilitate as fish sanctuaries. The transplantation of healthy stocks of desirable sport fishes have to be done on scientific lines on the basis of optimum biological productivity of the stream. The conservation and management of aquatic ecosystem of hill streams should involve the management of aquatic life including fishes and multiplicity of uses of a whole ecosystem as amenity, ecotourism etc.

Table 1.10: Fish Diversity in Sahastradhara Hill-stream at Dehradun

Scientific Name	*Local Name*
Schizothorax richardsonii	Asella
Schizothorax plagiostomus	Dhibrua
Schizothorax sinatus	Maseen
Puntius chola	—
Puntius hexasticus	Khront
Barilius barila	—
Barilius bendelisis	Chal
Barilius vagra	Chal
Noemacheilus savona	—
Glyptothorax pectinopterus	—
Mastacembelus armatus	—
Garra lamta	—
Garra gotyla gotyla	Gunthala
Danio devario	Patukari
Chagunia chagunio	Gelhari
Botia durlo	Baghaua
Chipisoma garua	Raikari
Colisa fasciatus	Khosti

Conclusion

Sahastradhara hill-stream ecosystem has suitable conditions for the habitation and survival of different biotic communities including plankton, nekton and benthos. In the Sahastradhara hill-stream Calcium and Magnesium level is high especially in the summer season. All the parameters studied had minor differences at different sites during the observation. Water quality of Sahastradhara stream has widely affected by tourists activities hence industrial pollution is no more here. Primary productivity is high in winter season whereas production efficiency was influenced by various physico-chemical parameters at different sampling sites.

The water quality is quite rich in natural nutrients which play an important implication for use the drinking purpose after making sure the bacterial and other hygenic condition after clean technology. The conservation measures must be directed towards understanding of the intricate relationship between all biotic organisms either plants or animals along with surrounding environment.

References

Ahamed, S. and Hasnain,S.I., 2001. Snow and stream water chemistry of the Ganga headwater basin, Garhwal Himalaya, India. *Hydrological Sciences Journal*, 46(1): 103–111.

Allen, W.E., 1920. A quantitative and statistical study of the plankton of the San Joaquin River and its tributaries in a near Stockton California in 1913. *Univ. Calif. Publ. Zool.*, 22(1): 1–24.

APHA, 1995. Ecological methods for the examination of water. *American Public Health Association*, 1015 fifteen street NW, Washington, 19: 1–1134.

Badola, S.P. and Singh, H.R., 1981. Hydrobiology of the river Alaknanda of Garhwal Himalaya *Indian J. Ecol.*, 8(2): 269–276.

Chopra, A.K. and Patrick, Nirmal, J., 1994. Effect of domestic sewage on self-purification of Ganga water at Rishikesh. *Bio Science*, 13(11): 75–82.

Das, S.M. and Pathani, S.S., 1978. A study on the effect of lake ecology on production of Mahaseer (*Tor tor* and *T. putitora*) in Kumaon lake. *Indian Matsya*, 4: 25–31.

Das, S.M. and Srivastava, V.K., 1956. Studies of fresh water plankton II correlation between plankton and hydro-biological factors. *Proc. Nat. Acad. Sci. India*, (13)26: 243–253.

Dobriyal, A.K., 1985. Ecology of limnofauna in the small streams and their importance to the village life in Garhwal Himalya. *Uttar Pradesh J. Zool.*, 5(2): 139–144.

Eddy, S., 1934. A study of freshwater plankton communities. *Illinois Biol. Monogr.*, 12(4): 1–93.

Hutchinson, G.C., 1957. *A Treatise on Limnology, Geography, Physical and Chemical*, Vol. 1. John Wiley and Sons, Inc., New York, pp. 1–1015.

Hynes, H.B.N., 1970. *The Ecology of Running Waters*. Liverpool University Press, Liverpool, 4th impression, p. 1–555.

Juttner, I., Heike Rothfritz and Steve J. Ormerod, 1995. *University of Wales, Cardiff, CFI 3TL UK*–'Diatom Communities in Himalayan hill streams, pp. 589–597.

Joshi, B.D., Bisht, R.C.S. and Semval, V.P., 1995. Primary productivity in western Ganga Canal at Hardwar. *Indian J. Ecol.* , 22(2): 123–126.

Kar, O.K., P.C. Mishra, M.C. Das and R.C. Das, 1978. Pollution studies in river Ib. III Plankton, pollution and primary productivity, p. 175.

Khan, A.A. and Siddiqui, A., 1971. *Hydrobiologia*, 37: 447.

Khanna, D.R., Badola, S.P., Sing, H.R. and Dobriyal, K., Anoop, 1992. Observations on seasonal trends in diatomic diversity in the river Ganga at Sapt-sarover, Hardwar. In: *Recent Research in Cold Water Fisheries*, (Ed.) K.L. Sehgal. Today and Tommorrow Printers and Publishers, New Delhi, p. 99–107.

Khanna, D.R., 1993. *Ecology and Pollution of Ganga River*. Ashish Publishing House, New Delhi, p. 1–241.

Khanna, D.R., Malik, D.S. and Rana, D.S., 1998. Phytoplanktonic communities in relation to certain physico-chemical parameters of Ganga canal at Hardwar. *Him. J. Env. Zool.*, 12: 193–197.

Khanna, D.R and Singh, R.K., 2000. Seasonal fluctuations in the plankton of Suswa River at Raiwala Dehradun. *Env. Conservations J.*, 1(2 and 3): 89–92.

Khanna, D.R., Bhutiani, R. and Chandra, K.S., 2005. Modeling to assess effect of light on phytoplanktonic growth dynamics: a review, *Env. Conservations J.*, 6(3): 103–107.

Lakshminarayana and R.K. Someshekar, 2001. *Ecology of Polluted Water,* (Ed.) Arvind Kumar Vol. 1, Chapter II, pp. 51–60. APH Pub. Corp. New Delhi.

Nautiyal, P., 1982. Some aspects of bioecology of Tarputitora in relation to hydrobiology of some Garhwal hill stream. *D.Phill. Thesis;* Garhwal University.

Pande, R.K. and Asha Mishra, 2000. Water quality study of freshwaters of Dehradun (Sahastradhara stream and Mussoorie lake). *Aquacult.,* (1): 57–62.

Quadri, M.Y. and Shah, G.M., 1984. Hydro-biological feature of Hoassar. A typical wetland of Kashmir–I, Biotope. *Indian J. Ecol.,* 2(2): 203–206.

Santhanam, R., Velayntham, P. and Jegatheesan, G., 1989. *A Manual of Freshwater Ecology.* Daya Publishing House, Delhi, p. 3–17, 53–92.

Sarin, M.M., Krishnaswamy, S.K., Trivedi, J.R. and Sharma, K.K., 1992. Major ion chemistry of the Ganga source waters: Weathering in high altitude Himalaya. In: *Proc. Indian Acad. Sci.* (*Earth Planet Sci.*) 1: 89–98.

Schwoerbel, J., 1991. *Handbook of Limnology.* Ellis Horwood Limited, Chichester, England, pp. 95–114.

Shrama, R.C., 1986. Effect of physico-chemical factors on benthic fauna of Bhargirathi River Garhwal Himalaya. *Indian. J. Ecol.,* 13(1): 133–137.

Singh, S.N., Gupta, M.L.and Prakash, P., 1980. Hydro-biological studies of some eutrophic ponds of Rohtash, Bihar. *Environment and Ecology,* 16(2): 457–462.

Subramanian, V., 1989. Chemical and suspended sediment characteristics of river of India. *J. Hydrol,* 44: 37–55.

Trivedi, R.K. and Goel, P.K., 1984. *Chemical and Biological Methods for Water Pollution Studies.* Environmental Publication, Karad, p.1–251.

Valdiya, K.S., 1975. Emergence of the Himalayas. *Science Today,* 9: 14–18.

Vashney, P.K., R. Vijayalakshmi, Nair and S.A.H. Abidi, 1983. Primary productivity in near shore water of Thai Maharashtra coast, India. *J.Mar. Sci.,* 12: 27–30.

Yeragi, S.G. and Nisar Shaikh, 2003. Studies on primary productivity of Tansa River. *J. Natcon,* 15(1): 125–130.

Chapter 2

Effects of Habitats on the Occurrence and Distribution of Blue Green Algae in North Maharashtra, India

S.R. Mahajan and S.N. Nandan

Introduction

Although a few attempts have been made in the past to study the algal flora from Jalgaon district from North Maharashtra, there appears to be no record as far as the distribution in different habitats is concerned. Thomas and Gonzalves (1965), Vasishta(1968), Barhate and Tarar(1983),Galankar (2000) have studied the algal flora from Unapdeo hot spring of this region. Kottawar *et al.* (1983), Mahajan (1986) explored the blue green algae from the cultivated banana fields of Jalgaon district. Barhate and Tarar (1983b), Bhoge and Ragothaman (1986) and Mahajan and Nandan (2006) have studied the blue green algae from the fresh waters in Jalgaon district.

The present investigation deals with the distribution pattern of blue green algae in three different habitats *viz.*, banana fields, fresh water and hot water spring in Jalgaon district. During this study a total of 51 taxa from banana fields, 122 from fresh waters and 55 taxa from the hot water spring have been recorded.

Three locations namely Jamner, Chopada and Yawal were selected for algal collection from banana fields. Fresh water algae were collected from Hartala lake of Muktainagar and Velhala lake of Bhusawal, while hot spring at Unapdeo of Chopada constituted the third habitat.

The algal collection were made from January 99 to December 2000 from the said location and were preserved in 4 per cent formaline. Soil samples from the banana fields have been regularly collected and cultivated in the laboratory in Bold's Basal medium (1942). The identification was done with the help of Desikachary (1959), some recent publications available were also consulted (Kamat 1963; Pandey and Mitra 1963; Prasad and Mehotra 1979 and Tiwari 1979).

Results and Discussion

In the present study, a total of 207 taxa of blue green algae have been recorded, of which 51 belongs to Chroococcaceae, 91 to Oscillatoriaceae, 33 to Nostocaceae, 08 to Scytonemataceae and 24 to Rivulariaceae (Table 2.1).

Table 2.1: List of Blue Green Algae from Different Families

Chroococcaceae	*Microcystis aeruginosa, M. flos-aqueae, M. holsatica, M. marginata, M. protocystis, M. robusta, M. viridis, Chroococcus cohaerens, C. hansgirgi, C. micrococcus, C. minimus, C. minutus, C. montanus, C. tenax, C. turgidus, C. varius, Gloeocapsa attrata, G. calcarea, G. kuetzingiana, G. crepidinum, G. fuscolutea, G. gelatinosa, G. punctata, G. stegophila, G. stegophila v. crassa, Gloeothece membranacea, G. rupestris v. maxima, G. samoensis, G. samoensis v. major, Aphanocapsa banaresensis, A. bullosa, A. crassa, A. grevillei, A. koordersi, A. Montana, A. pulchra, A. roseana, Aphanothece castagne, A. pallida, A. saxicola, A. stagnina, Synechococcusa ruginosus, S. elongates, Synechocystis aqualitis, S. pevalekii, Coelosphaerium kuetzingianuml, Merismospedia, convoluta, M. elagans, M. minima, M. punctata, M. tenuissima,*
Oscillatoriaceae	*Spirulina abbreviate, S. gigantean, S. major, S. meneghiniana, Oscillatoria accutissimum, O. agardhi, O. amoena, O. brevis, O. calcuttensi, O. chilkensis. O. curviceps v. angusta, O. chalybea, O. chlorine, O. formosa f. loktakensis, O. hamelli, O. Irrigua, O. jasorvensis, O. laete-virens, O. limnetica, O. limosa, O. margaretifera, O. maharashtrensis, O. nigro-virdis. O. obscura, O. ornate, O. profunda, O. perornatra, O. perornata v. attenuata O. princeps, O. pseudogeminata, O. pseudogeminata v. unigranulata, O. quadripanctulata, O. raoi, O. raciborskii, O. rubescens, O. subtillissima, O. schultzii, O. simplicissima, O. subuliformis O. terebriformis, O. tenuis, O. tenuis g. v. tergestina, O. vizagapatensis, O. willei, Phormidium ambiguum, P. angustissimum, P. autumnale, P. bigranulatum, P. calcicola, P. corium v. capitatum, P. foeveloarum, P. fragile, P. gyralis, P. incrustatum, P. inundatum, P. jadinianum, P. lucicium, P. mucosum, P. pachydermaticum, P. papyraceum, P. purpurascens, P. retzii, P. rotheanum. P. subincrustatum, P. tenue, P. usterii, Lyngbya aestuarii, L. allorgei, L. borgertii, L. bacculum, L. ceylanicav. constricta, L. cryptovaginata, L. digueti, L. epiphytica L. gracilis, L. holsatica, L. hieronymusii, L. hieronymusii. v crassivaginata, L. infixa, L. laxespiralis, L. limnetica, L. martensiana, L. mesotricha, L. nordgardhii, L. perelegans, L. putealis Mont, Schizothrix muelleri, Symploca cartilaginea, S. muralis, S. muscorum, Microcoleus lacustris.*
Nostocaceae	*Cylindrospermum indicum, C. muscicola, C. plantonicum, C. stagnale, Nostoc calcicola, N. carneum, N. coeruleum, N. ellipsosporum, N. linckia, N. muscorum, N. piscinale, N. pruniformae, N. palludosum, N. punctiformae, N. sphaericum, N. spongiaeformae v. varians, Anabaena ambigua, A. amomala, A. doleolum, A. fertilisimma, A. iyengarii, A. khannae, A. laxa, A. naviculoides, A. oryzae, A. orientalis, A. sphaerica, A. spinosa, A. spiroides, A. variabilis, Raphidiopsis indica, R. curvata, R. mediterranea.*
Scytonemataceae	*Plectonema notatum, Scytonema bewsii, S. coactile, S. fritschii, Tolypothrix distorta, T. penicillate, T. nodosa, T. robusta.*
Rivulariaceae	*Homoeothrix balearica H. Juliana, Calothrix brevissima, C. castelii, C. clavata, C. desikacharyensis C. elenkinii, C. fusca, C. javanica, C. marchica, C. membranicea, C. perientina, C. thermalis, Rivularia aquatica, R. beccariana, R. globiceps, Gloeotrichia ghosei, G. indica, G. pilgeri, G. raciborskii, G. raciborskii v. bombayense, G. raciborskii v. conica, G. raciborskii v. kashiense, G. raciborskii v. longispora*

Majority of the species recorded belong to Nostocales in which members of Oscillatoriaceae formed a major group followed by the members of Chroococcaceae.

A comparative account of the work done on blue green algae from different habitats in Jalgaon district has been shown in Table 2.2.

Table 2.2: Work Done on Blue Green Algae from Different Habitat

Authors	*Freshwater Habitats*	*Hot Water Habitats*	*Banana Fields*
Thomas and Gonzalves (1965)		Chroococcaceae (07)	
		Oscillatoriaceae (14)	
Vasishta (1968)		Chroococcaceae (03)	
		Oscillatoriaceae (10)	
Barhate and Tarar	Chroococcaceae (22)	Chroococcaceae (15)	
(1983a) (1983b)	Oscillatoriaceae (50)	Oscillatoriaceae (32)	
	Nostoacaceae (1)	Nostoacaceae (3)	
	Stigonenataceae (1)		
Kottawar *et al.* (1983)			Chroococcaceae (03)
			Oscillatoriaceae (08)
			Nostoacaceae (04)
			Rivulariaceae (01)
Bhoge and Ragothamn (1986)	Chroococcaceae (26)		
	Oscillatoriaceae (31)		
	Nostoacaceae (06)		
Mahajan and Mahajan (1987)	Chroococcaceae (09)	Chroococcaceae (20)	Chroococcaceae (12)
	Oscillatoriaceae (22)	Oscillatoriaceae (20)	Oscillatoriaceae (21)
	Nostoacaceae (15)	Nostoacaceae (05)	Nostoacaceae (14)
	Scytonemataceae (06)		Rivulariaceae (11)
	Rivulariaceae (08)		
Galankar (2000)		Chroococcaceae (20)	
		Oscillatoriaceae (43)	
		Nostoacaceae (02)	
Present investigation	Chroococcaceae (22)	Chroococcaceae (18)	Chroococcaceae (13)
	Oscillatoriaceae (55)	Oscillatoriaceae (33)	Oscillatoriaceae (22)
	Nostoacaceae (14)	Nostoacaceae (04)	Nostoacaceae (15)
	Scytonemataceae (07)		Rivulariaceae (15)
	Rivulariaceae (09)		

Table 2.3: Range of Temperature and pH Recorded by Different Authors at Unapdev Hot Spring

Authors	*Temperature*	*pH*
Thomas and Gonzalves (1963)	39–60°C	–
Vasishta (1968)	54.4–60°C	8.5
Barhate and Tarar (1983a)	40–50°C	7.5
Mahajan and Mahajan (1987)	40–60°C	7–9
Galankar (2000)	54.5–67°C	7.3–8.6
Present investigation	54–65°C	7.5–8.5

It is observed that 14 species belong to Oscillatoriaceae is common in both fresh water as well as banana field habitats.

From the study of comparative account of algae in banana fields, it is seen that results obtained in the present study agree well with those observed by Kottawar *et al.* (1985) except the total number of taxa recorded. In the present study a total of 65 taxa have been obtained from the cultures of banana field soil from three different localities, while Kottawar *et al.* (1985) have studied the blue green algae from the culture of banana field soil from one locality only.

The number of algal taxa recorded from the fresh water habitat also agree with the account given by Barhate and Tarar (1983) and Bhoge and Ragothaman (1986). Increased in the number of Nostocaceae members have been recorded during the present study. It might be due to high contents of nitrates present in water of Hartala lake and Velhala lake

As far as the hot water habitats is concerned the results obtained are similar with those of Barhate and Tarar (1983a) and Galankar (2000). Thomas and Gonzalves (1965) and Vasishtha (1968) have recorded comparatively less number of taxa but there also predominance of Oscillatoriaceae members is observed. Thus the results of present study also tally with those of the earlier workers, the difference is being the total number of taxa. Thus increase might have taken place due to adjustment of same additional blue green algal taxa in the hot water habitats.

Range of temperature and pH of the hot water recorded has been shown in Table 2.3. It has been observed that maximum algal taxa have been recorded between 54.5°C–67°C. From the study of pH range, it is seen that the water is alkaline with pH ranging between 7.5-8.5.

Acknowledgements

The authors are thankful to Dr.(Mrs.) Meenakshi Waykole, Principal, Bhusawal Arts, Science and P.O.N Commerce College, Bhusawal for providing the laboratory facilities.

References

Barhate, V.P. and Tarar, J.L., 1983. Thermal algae of Khandesh, Maharashtra. *Trop. Plant Science Research,* 1(2): 195–196.

Barhate, V.P. and Tarar, J.L., 1983. Algae of Maharashtra, Cyanophyceae of Khandesh-I. *Phykos,* 22: 67–72.

Bhoge, O.N. and Ragothaman, G., 1986. Studies on the Cyanophyceae from region, Maharashtra. *Phykos,* 25(1/2): 129–131.

Bold, H.C., 1942. The culitivation of algae. *Bot. Rev.,* 8: 69–138.

Desikachary, T.V., 1959. *Cyanophyta.* I.C.A.R. New Delhi

Galankar,V.S., 2000. Limnology of algae of certaion thermal springs of Maharashtra. *Ph.D.Thesis,* North Maharashtra University, Jalgaon.

Kamat, N.D, 1963. The algae of Kolhapur, India. *Hydrobiol.,* 22(3/4): 209–309.

Kottawar, S.T., Pachpande, R.R. and David, S.B., 1983. Soil algae of banana fields of Jalgaon District (M.S.). *Proc. A.I.A.P.C.,* Kanpur, p. 211–212.

Mahajan, A.D., 1986. Some *Oscillatoria* Vaucher and *Phormidium* Kuetz. from banana field of Jalgaon District, Maharashtra. *Indian Bot. Reptr.,* 5(1): 42–26.

Mahajan A.D. and Neelima Mahajan, 1987. Habitats effect on the occurrence and distribution of blue green algae in Jalgaon District (M.S). In: *Recent Researches in Ecology Environment and Pollution*, 5: 151–159.

Mahajan, S.R. and S.N. Nandan, 2006. Studies on algae of polluted lakes of Jalgaon (Maharashtra):Role of blue green algae. In: *Ecology of Lakes and Reservoirs*, (Ed.) V.B. Sakhare, pp. 54–62. Daya Publishing House, New Delhi, India.

Pandey, D.C. and Mitra, A.K., 1963. On the morphology and life history of new species of *Anabaenopsis*, *A. ambiqua*. n.sp. *Nova Hedwigia*, 4(3/4): 345–350.

Prasad, B.N. and Mehrotra, R.K., 1976. *Calothrix lucknowenes*, a new species from paddy field soils. *Phykos*, 15(1/2): 67–73.

Prasad, B.N. Mehrotra, R.K. and Singh, Y., 1977. Some new taxa of *Nostoc* vaucher from North Indian crop fields. *Geophytology*, 7(2): 222–228.

Thomas, J. and Gonzalves, E.A., 1965. Thermal algae of western Hot springs at Unai, Lasundra, Unapdeo. *Hydrobiol.*, 26(1/2): 55–65.

Tiwari, G.L., 1979. A study of blue green algae for Paddy field soils of India Part-3.Taxonomic considerations of Nostocales and Stigonematales. *Nova Hedwigia*, 63: 139–159.

Vasishta, P.C., 1968. Thermal cyanophyceae of India part-I. *Phykos*, 7(1/2): 198–241.

Chapter 3

Zooplankton Composition in Vellar Estuary in Relation to Shrimp Farming

M. Rajasegar, M. Srinivasan and S. Ajmal Khan

Introduction

Shrimp farming picked up fast in India particularly in the southern states, due to its promotion on the part of government agencies. The discharge of used water from the farms into the estuaries, backwaters and mangroves became a matter of great concern to the environmentalist. In the absence of reliable data collected over a period of time, no definite conclusion could be drawn about the negative impact of shrimp farming. This emphasized the need for monitoring different physico-chemical and biological parameters of the estuaries, backwaters and mangroves in the vicinity of shrimp farms. Shrimp farms started their operation in and around Vellar estuary (Lat.11°29′ N; 79°46′ E) in the year 1992 and by 1995, there were as many as 42 shrimp farms with water spread area of 150 ha. The formulated shrimp feed pellets used are rich in protein, fat, fiber and minerals etc. These nutrients get dissolved in water, settle in the sediments of the shrimp farms and are discharged into the Vellar estuary during periodic draining. Therefore the present study was undertaken to ascertaining changes in the biota arising out of discharge of used water from shrimp farms into the estuary.

Materials and Methods

Surface water and zooplankton samples were collected at monthly interval from three stations (Figure 3.1) during January 1996–December 1997, using a net (mouth area $0.25m^2$ and mesh width of 120µm). The plankton samples were preserved in 4 per cent buffered formaldehyde in seawater immediately after their collection. Species diversity index (H′) was calculated using Shannon and Wiener's (1949) formula, species richness was calculated as described by Gleason (1942) and evenness using the formula of Pielou (1966). Hydrographical parameter such as temperature was measured using a standard Celsius thermometer, salinity was estimated with the help of a salinometer model E-2 and pH was measured using a Elico pH meter. Dissolved oxygen was determined by using standard method (Strickland and Parson, 1972).

Results and Discussion

The range and mean value of physico-chemical parameters of surface water and zooplankton parameters are given in the Table 3.1. Variations in water temperature were mainly influenced by the atmospheric temperature. pH varied from 7.1 to 8.5 with higher values during summer season and low values during monsoon. During summer period salinity increased to 36‰ due to low fresh water inflow and drastically decreased to 0 ‰ during monsoon due to high fresh water influx. The dissolved oxygen values varied between 4.0 and 7.5ml/l. High levels of total phosphate, inorganic phosphate and nitrate recorded during monsoon season due to fresh water inflow and washing off of fertilizers used in the agriculture fields and also discharge from shrimp farms and low values during summer season due to decreased fresh water inflow and utilization by phytoplankton.

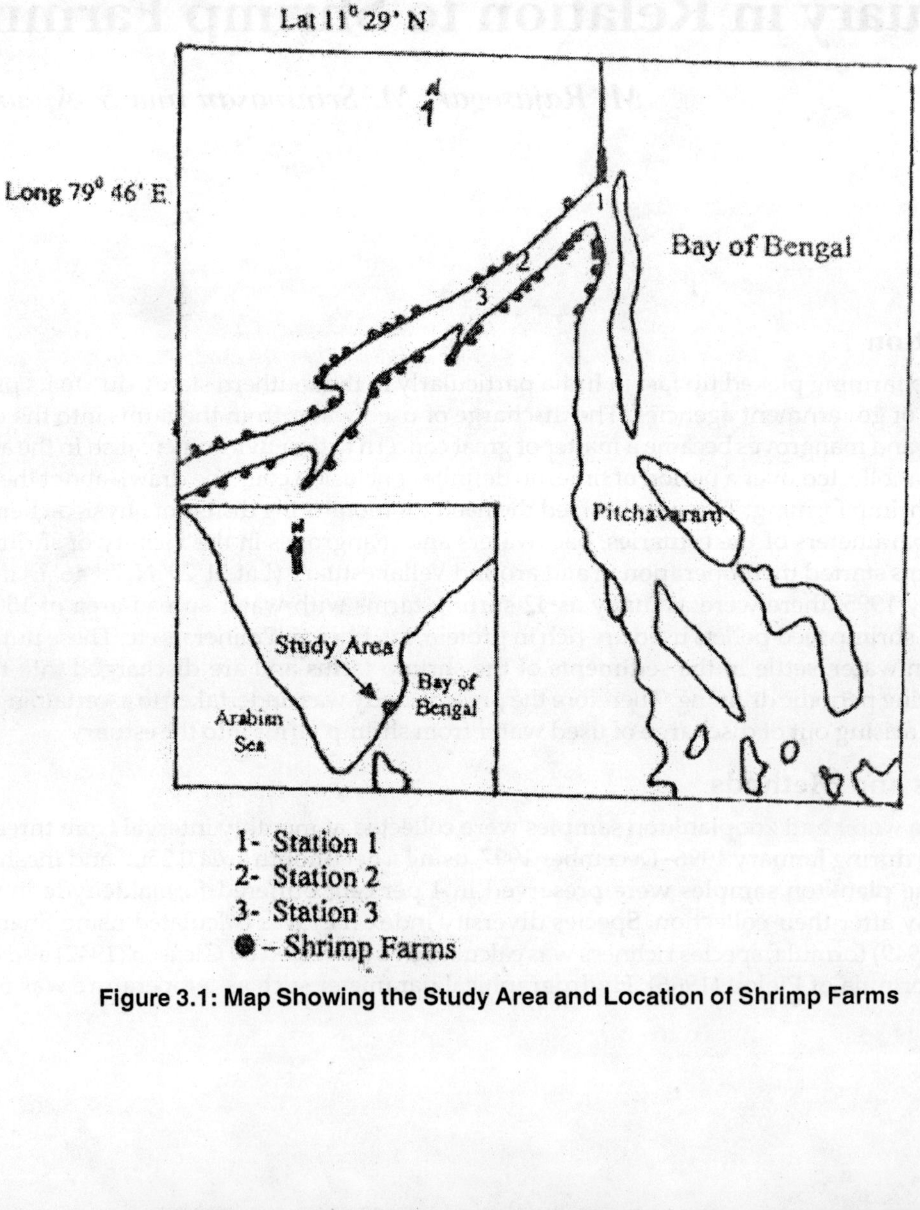

Figure 3.1: Map Showing the Study Area and Location of Shrimp Farms

Table 3.1: Annual Range and Mean Values of Physico-chemical and Zooplankton Parameters at Stations 1, 2 and 3

Parameter	*Year*	*Station 1*			*Station 2*			*Station 3*		
		Min.	*Max.*	*Mean*	*Min.*	*Max.*	*Mean*	*Min.*	*Max.*	*Mean*
Surface water temperature (°C)	1996	23.0	34.5	29.5	23.5	34.5	29.7	23.0	35.0	29.7
	1997	23.5	33.5	28.9	23.5	34.0	29.0	23.0	34.5	29.0
pH	1996	7.4	8.5	7.9	7.4	8.4	7.8	7.4	8.1	7.6
	1997	7.3	8.4	8.0	7.1	8.3	7.9	7.1	8.1	7.8
Salinity (per cent)	1996	6	36	28	6	36	26	4	34	23
	1997	4	36	27	4	36	25	0	35	24
Dissolved oxygen (ml/l)	1996	4.3	7.4	6.4	4.2	7.3	6.2	4.0	7.4	6.1
	1997	4.5	7.5	6.6	4.2	7.1	6.4	4.0	7.3	6.2
Total phosphorus (µM)	1996	0.34	2.17	0.94	0.42	2.12	0.93	0.38	2.15	0.93
	1997	0.24	2.23	0.92	0.19	2.21	0.84	0.23	2.19	0.88
Inorganic phosphate (µM)	1996	0.13	1.78	0.60	0.09	1.74	0.59	0.17	1.89	0.61
	1997	0.16	1.83	0.66	0.21	1.77	0.65	0.16	1.87	0.57
Nitrate (µM)	1996	0.13	2.32	1.02	0.10	2.30	0.98	0.09	2.40	0.98
	1997	0.08	2.34	0.85	0.08	2.31	0.81	0.06	2.34	0.80
Zooplankton density (no/m^3)	1996	220	62000	29120	240	64500	29850	180	63500	29270
	1997	300	74500	32175	260	77000	32570	200	75100	31860
Species diversity (bits/ind.)	1996	1.57	3.62	2.90	1.37	3.59	2.82	1.17	3.66	2.72
	1997	1.32	3.56	2.74	1.21	3.49	2.61	0.91	3.47	2.53
Species richness	1996	0.91	2.92	2.02	0.90	2.86	1.98	0.83	2.91	1.99
	1997	0.83	2.99	2.03	0.82	2.93	2.03	0.71	2.88	1.99
Species evenness	1996	0.68	0.98	0.90	0.69	0.99	0.90	0.67	0.99	0.89
	1997	0.72	0.99	0.90	0.73	0.99	0.91	0.71	0.98	0.89

Abundance of various zooplankton groups in the estuary fluctuated in accordance with salinity regime and copepods formed the largest component. There were 32 species, calanoids constituted the bulk of copepods followed by cyclopoids and harpacticoids. The most common species of copepods were *Acartia danae, A. erythreae, A. southwelli, A.spinicauda, Oithona brevicornis, O.plumifera, O.similes, O.rigida, Pontella danae, P.securifer, P.spinipes, Pseudodiaptomus annadalei, P.aurivilli, P.serricaudatus, Temora discaudata, T.stylifer* and *T.turbinata.*

Among the ciliates *Tintinnopsis* was the most abundant genus with 7 species (*T.beroidea, T.butschelli, T.cylindrica, T.directa, T.minuta, T.gracilis* and *T.tubulosa*). These tolerate wide variation in salinity but higher numbers were recorded during summer, which might be due to the influence of neritic waters in the estuary.

Among the rotifers, *Brachionus* was the most abundant genus with 3 species (*B.calyciflorus, B.caudatus* and *B.rubens*). Rotifers were observed only during the monsoon and post monsoon seasons as they are not tolerant to higher salinity. Similar type of observations were made from Vellar estuary (Chandran, 1982), Coleroon estuary (Patterson Edward, 1988), Pichavaram mangroves (Karuppasamy and Perumal, 2000) and Pondicherry coast (Ananthan, 1994).

Among decapods, only one species *Lucifer hansenii* was recorded at all stations and chaetognaths such as *Sagitta* sp. was abundant throughout the study period. But during monsoon months when fresh water dominated in the estuary, it was absent. This is confirmation with the earlier findings from Vellar estuary (Chandran, 1982; Thangaraj, 1984).

Among the cladocerans, *Evadne tergestina* and *Penilia avirostris* were the most common and occurred throughout the study period. Among the larvae, veliger, zoea, megalopa, ciripede nauplius, copepods nauplii, gastropod nauplii, hermit crab larvae, mysis larvae, oikeopleura and echinodermata larvae were abundant. Fish eggs and larvae were commonly found at all the stations in large numbers.

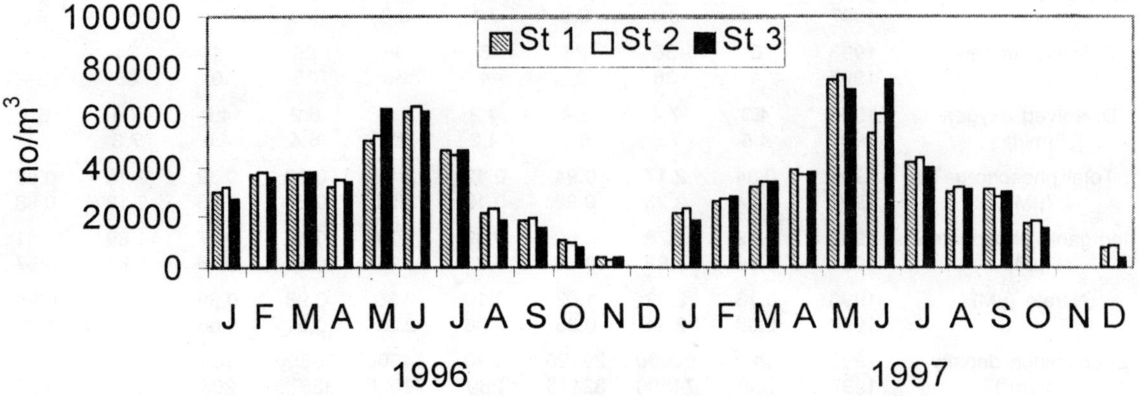

Figure 3.2: Monthly Variations in Zooplankton Population Density Recorded from Stations 1, 2 and 3

The monthly variations in density (Figure 3.2), diversity (Figure 3.3), richness (Figure 3.4) and evenness (Figure 3.5) with range and mean values are given in Table 3.1.The maximum zooplankton density was observed during the summer season and minimum during the monsoon season. The high values in summer could be due to relatively stable environmental conditions, and the neritic influence

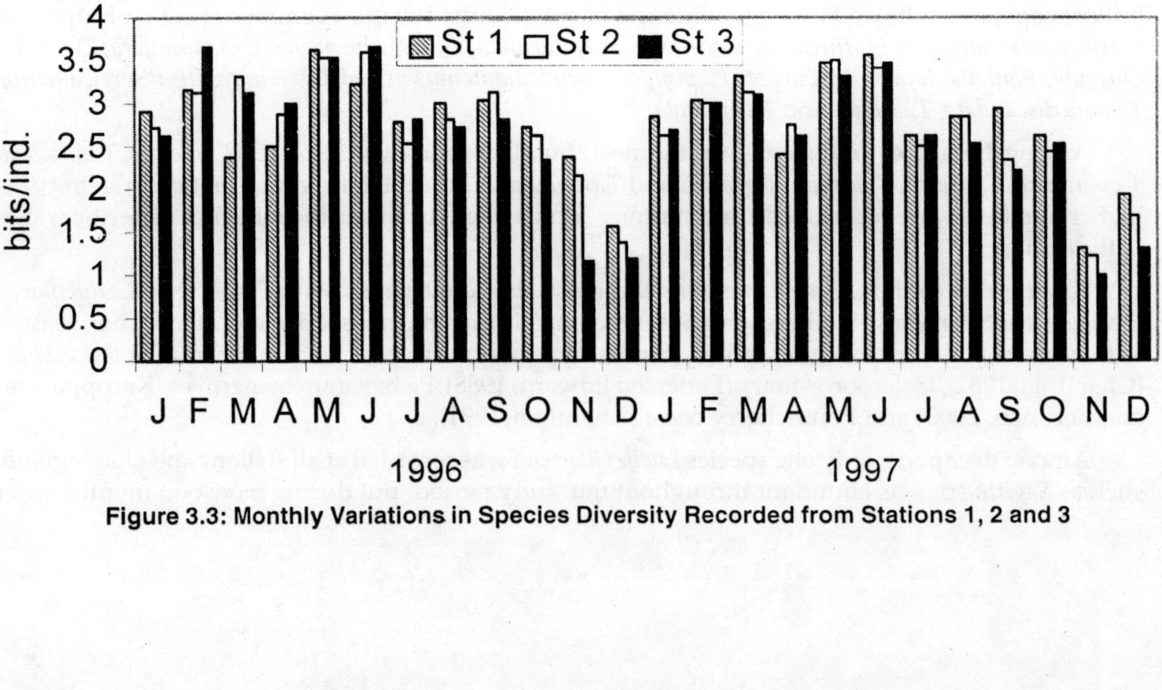

Figure 3.3: Monthly Variations in Species Diversity Recorded from Stations 1, 2 and 3

in the absence of fresh water inflow. The salinity showed positive correlation with zooplankton density (r= 0.75, P<0.001 at station1; r=0.84, P<0.001 at station 2; r=0.83, P<0.001 at station 3). Zooplankton population density was low during monsoon season as the fresh water inflow changed the salinity, temperature and other environmental variables which inturn decreased the zooplankton density (Padmavathi and Goswami, 1996).

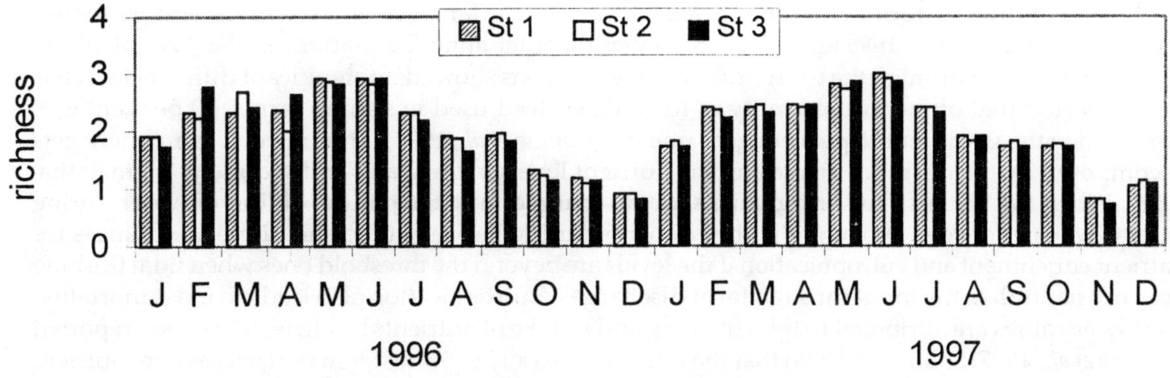

Figure 3.4: Monthly Variations in Species Richness Recorded from Stations 1, 2 and 3

Figure 3.5: Monthly Variations in Species Evenness Recorded from Stations 1, 2 and 3

Maximum species diversity, richness and evenness of zooplankton were recorded during summer season. The low species diversity was observed during the monsoon season. The diversity showed significant positive correlation with density ($r=0.74$, $P<0.001$ at station 1; $r=0.79$, $P<0.001$ at station 2; $r=0.80$, $P<0.001$ at station 3). The population density, species diversity, richness and evenness recorded in the present study are comparable to the earlier observations made in this estuary (Sivakumar, 1982; Kumar, 1991).

The quantities of formulated feed used in the shrimp farms here were 117.4 and 90 tonnes respectively in the years 1996 and 1997. However, the inter annual variations in the 3 zooplankton composition tested through the t-test in all the three stations showed no significant differences. It has been reported that of the total quantity of formulated feed used in shrimp farms, 10 per cent gets dissolved in the water and 15 per cent goes uneaten (Diana *et al.*, 1994). This uneaten component gets decomposed by the microbes increasing the nutrient levels in the water and sediment (Hemalatha, 1996). When the water from shrimp farms are discharged in to the natural bodies of water during regular water exchange and along with sediments during harvest at the end, there are chances for nutrient enrichment and eutrophication if the levels are beyond the threshold ones when tidal flushing does not neutralize the input through farm discharges. Eutrophication also leads to mass mortality. The lower values are attributed to tidal flushing and uptake of nutrients by algae. It has been reported (Philips *et al.*, 1997: Sakthivel, 1998) that the oysters generally grow faster in estuaries where nutrient levels are elevated. It is borne by the fact that the settlement of spats and profuse growth of oyster colonies have increased in the estuary recently. However during the present study mass mortality was not observed. Thus this study clearly showed that the discharges from the shrimp farms did not cause any significant and irreversible changes in the zooplankton composition. When the used water in shrimp farms is discharged after treatment as stipulated by the Aquaculture Regulation Act, the influence will be minimal and estuarine environments can be utilized for shrimp farming up to the carrying capacity of each eco-system.

Summary

There are as many as 42 shrimp farms situated on the banks of Vellar estuary. These farms discharge the used water into the estuary, which may influence the biota there. In the present study the distribution and abundance of zooplankton in relation to shrimp farming were studied in 3 stations of the estuary. Zooplankton density, diversity and richness were observed to be higher during the summer season and low during monsoon due to influence of hydrographic parameters. Abundance of various zooplankton groups in the estuary fluctuated in accordance with salinity regime and copepods formed the largest component. The maximum zooplankton density was observed during the summer season and minimum during the monsoon season. The high values in summer could be due to relatively stable environmental conditions, and the neritic influence in the absence of fresh water inflow. Maximum species diversity, richness and evenness of zooplankton were recorded during summer season. The low species diversity was observed during the monsoon season. When compared with the previous data from Vellar estuary there was no much difference in zooplankton diversity due to shrimp farming.

Acknowledgements

We wish to thank the Director of Centre for providing facilities. The first author thanks to Dr.A.Yogamoorthi, Centre for Future Studies, Pondicherry University for his valuable suggestions. The help and support rendered by the owners of shrimp farms is gratefully acknowledged.

References

Ananthan, G., 1994. Plankton ecology and heavy metal studies in the marine environs of Pondicherry, India. *Ph.D. Thesis*, Annamalai University, India, p. 125.

Chandran, R., 1982.Hydrobiological studies in the gradient zone of Vellar estuary. *Ph.D. Thesis*, Annamalai University, India, p. 195.

Diana, T.S., C. Kwei-lin and K. Jaiyen, 1994. Supplemental feeding of *Tilapia* in fertilized ponds. *J. World Aquacult. Soc.*, 25(4): 497–506.

Gleason, H.A., 1922. On the relation between species and area. *Ecology*, 3: 156–162.

Hemalatha, A., 1996. Studies on the benthos from a shrimp farm and in a Vellar estuary, Parangipettai. *M.Phil. Thesis*, Annamalai University, p. 39.

Karuppasamy, P.K. and P. Perumal, 2000. Biodiversity of zooplankton at Pichavaram mangroves, South India. *Ad. Bios.*, 19(2): 23–32.

Padmavathi, G. and S.C. Goswami, 1996. Zooplankton ecology in the Mandovi-Zuari estuarine system of Goa, west coast of India. *Indian J. Mar. Sci.*, 25: 268–273.

Paterson Edward, J. K., 1988. Studies on the ecology of Coleroon estuary, southeast of India. *Ph.D. Thesis*, Annamalai University, India, p. 228.

Philips, J., M.D.J. Macintosh and Pedro Bueno, 1997. Environmental issues relating to aquaculture. *MPEDA Newsletter*, 2(4): 13–25.

Pielou, E.C., 1966. The measurement of diversity in different type of biological collections. *J. Theoret. Biol.*, 13: 144.

Sakthivel, M., 1998. Aquaculture: Food gold of the future. *The Hindu Survey of the Environment*, p. 35-39.

Shannon, C.E. and W. Wiener's, 1949. *The Mathematical Theory of Communication*. University of Illinois press, Urbana, 117 pp.

Strickland, J.D.H. and T.R. Parsons, 1972. A practical handbook of seawater analysis. *Bulletin of Fisheries Research Board Canada*, 167: 310.

Thangaraj,G.S., 1984. Ecobiology of the marine zone of the Vellar estuary. *Ph.D. Thesis*, Annamalai University, India, p. 194.

Chapter 4

Kinetic of Planktons in Integrated Fish-Livestock Ponds Situated at Mid-Hill Altitude of Meghalaya, North Eastern India

Sullip K. Majhi

Introduction

The planktons are widely distributed in the aquatic environments and are most preferred food items for the cultured fish. However, the acceptability of plankton as a live feed by fish varies according to the species and feeding habits. In addition to this, planktons are also a good source of amino acid and other micro nutrients that helps in boosting the immune system of the fish, besides adding to the flesh biomass. In recent years, integrated fish farming has been gaining considerable importance in Meghalaya. Many farmers have begun switching from traditional method to integrated fish farming as a means of produce safe foodstuff and respecting environment. The integrated farming system favors lower input costs, conserve nonrenewable resources, besides improving quality of the product. The North Eastern Hill Region of India has been identified as the potential zone for integrated fish farming due to huge population of domestic livestock and negligible use of inorganic inputs. At present, aquaculture is one of the important enterprises in the state of Meghalaya and many rural farmers in valley areas have been found to have at least one or two ponds within their farm area for fish culture (Majhi, 2005). Furthermore, this part of the country receives rainfall during most part of the year, thus giving ample scope to fish culture throughout the year. Having knowing these facts, we at ICAR Research Complex, Barapani have been disseminating the technology of Integrated Fish Farming (IFF) to the farming community of the region for over past 15 years. However, very recently it has been realized to study the plankton populations, both qualitative and quantitative in different integrated units. In that front, the present study is a preliminary investigation to the plankton dynamics in an integrated fish-livestock pond situated at mid-hill altitude. Nevertheless, the baseline information generated from this study would go a long way in helping the farming community of the region by educating them on production of different phyto and zooplanktons that are mostly preferred by the cultured fish species under the integrated farming mode, thus making the integrated farming system more meaningful and profitable in the North Eastern Hill Region of India. Under this backdrop, the present study was conducted to gather baseline information on qualitative and quantitative plankton production by using livestock manures of poultry, pig and duck.

Materials and Methods

Three ponds of 1000m^2 each were used for the study. The ponds were located at 950 m above the MSL between 21.50°–29.50°N latitude and 85.5°–97.5°E longitude, inside the campus of ICAR Research Complex, Meghalaya. The climatic condition of the region is cold to warm pre-humid with annual rainfall of above 2000mm. The air and water temperature varies between 11–26°C and 10–24°C, respectively. Before the experiment was conducted, the ponds were kept fallow for over two months as recession period. After that, the ponds were dried and the bottom soil was exposed to sun for one week for eradication of unwanted microorganisms and wild fish. The manures of poultry (T1), pig (T2) and duck (T3) were applied at the rate of 1 Kg/m^2 to the respective ponds for fertilization. Auto filling of water was performed in each pond through underground seepage, which took one week to reach a water table of 1.0 m. The pond was kept without stocking of fish for 14 days for production of planktons. However, on 15 days onwards, the livestock manures were applied to the respective ponds @ 10 per cent kg body weight of the standing fish biomass. The experiment was conducted for a period of 180 days from March–August 2005.

The water quality parameters (variations in water temperature, pH, dissolved oxygen and total alkalinity) of the experimental ponds were monitored monthly and the samples were analyzed following the standard methods (APHA,1998).

Sample Preparations and Analysis

The plankton samples were collected every month from each pond by filtering 50 liters of water on standard plankton net. The live samples were brought to the laboratory for preliminary observations and then were fixed at 5 per cent formalin and stored in glass bottles for detailed qualitative and quantitative estimation of plankton and the results have been presented based on the pooled data of the samplings. For proximal analysis, livestock manures of poultry, pig and duck were collected and kept inside the oven (Universal made) at 60°C for 12 hours. Then after, the samples were removed and kept in the desiccators until the sample reaches ambient temperature. This protocol for drying was repeated until the sample attained a constant weight. The final weights of the sample were taken in digital balance (Mettler AB104-S) and were grinded in to powder form for further analysis.

Results

The variations in water temperature, pH, dissolved oxygen and total alkalinity during the experiment were 23–26°C, 7.0–8.0, 6.0–7.5mg/l (T3) and 43 to 45 ppm, respectively. The phytoplankton and zooplankton recorded from three different livestock manure treated ponds (T1, T2 and T3) are presented in Table 4.1, 4.2. The data presented are pooled data from the monthly sampling. At the end of the experiment, it has been observed that, among phytoplankton, the population of *Crucigenia sp* was more in T1 and *Kirchneriella sp* was found to be dominant genera in T2. Similarly, in the T3 which was treated with duck manure, *Sphaerocystis sp* was found to be quantitatively dominant phytoplankton genera. So far as the zooplankton populations are concerned, in T1, *Cyclops sp* was found to be the most dominant genera. At the same time, in T2 and T3, *Nauplius sp* was found to be the most dominant zooplankton genera. Further, during the present study, seasonal food competition among the planktonic populations were also observed, by virtue of which, some genera of phytoplankton and zooplankton had attended pick during certain samplings. However, in the current article those data are not discussed and it needs further validation before making any concluding remarks. In addition to this, the proximal analysis of the livestock manures revealed that, among the three, poultry manure was found to be containing more micronutrients (3.20 per cent N: 2.20 per cent P: 1.10 per cent K), followed by duck manure and pig manure (Table 4.3). Overall, the quantitative phytoplankton production was highest in T3 (31640/lt) > T1 (27878/lt) >T2 (20671/lt) and zooplanktons in T3 (15117/lt) >T1 (12978/lt) >T2 (4964/lt). In addition to this, there were certain genera of phytoplankton and zooplanktons which were exclusively found in T3 and were not observed in T2 and T1, signifying the rich plankton diversity in T3. Thus, in the present experiment it has been concluded that duck manure is the best for both qualitative and quantitative plankton production under Integrated Fish Farming followed by poultry manure and pig manure.

Table 4.1: Qualitative Phytoplankton Distribution in Three Treatment Ponds

Genera	*Poultry Manure (T1)*	*Pig Manure (T2)*	*Duck Manure (T3)*
Botryococcus sp	1810	1200	2020
Selenastrum sp	1750	1000	1016
Euglena sp	1210	1004	2001
Hydrodictyon sp	1002	1001	1000
Ciliates sp	1590	1100	2090
Kirchneriella sp	1620	1710	3410
Nostoc sp	900	920	1000
Crucigenia sp	1900	1000	990
Sphaerocystis sp	1660	1000	1100
Filamentious sp	500	600	910
Scenedesmus sp	1220	1004	1003
Phacus sp	1004	1002	1001
Cosmerium sp	1260	0	1008
Navicula sp	1012	1008	1000
Ankistrodesmus sp	1006	1000	1004
Ceratium sp	1012	1010	1020
Pediastrum sp	1014	1008	1005
Anthonema sp	1006	0	1300
Clostridium sp	1002	0	1008
Xanthidium sp	1100	1000	1010
Fragillaria sp	1000	1004	1015
Sphaerocystis sp	1000	1100	3720
Trichocerca sp	1300	1000	1009
Total	**27878**	**20671**	**31640**

Table 4.2: Qualitative Zooplankton Distribution in Three Treatment Ponds

Genera	*Poultry Manure (T1)*	*Pig Manure (T2)*	*Duck Manure (T3)*
Notholca sp	1001	1008	900
Selenastrum sp	1058	200	800
Epbemera sp	1008	100	950
Pompbolyx sp	1022	100	601
Calanoid sp	1003	1002	1004
Eschaneustyla sp	0	0	1005
Polyarthra sp	980	100	970
Pbilodina sp	1006	200	1020
Cyclops sp	1100	1000	990
Lecane sp	1010	0	1001
Gastropus sp	0	0	1200
Keratella sp	0	0	1100
Nauplius sp	1009	1104	1206
Monostyla sp	750	0	600
Macrothrix sp	1020	100	850
Brachionus sp	1011	50	920
Total	**12978**	**4964**	**15117**

Table 4.3: Chemical Composition of Livestock Manure (Dry Weight Basis)

Manures	*Moisture*	*Chemical Composition (Per cent of Dry Matter)*		
		Nitrogen	*Phosphate*	*Potassium*
Pig manure	67-69	1.35	0.37	0.28
Duck manure	81-83	0.93	0.40	0.38
Poultry manure	17-20	3.20	2.20	1.10

Discussion

The integrated livestock fish farming system is a proven environmentally sustainable and economically viable technology that encompasses rational utilization of available resources and livestock such as duck, poultry and pigs (NACA, 1989). Over the years, different forms of integrated livestock fish farming *viz.*, duck-fish, poultry-fish, pig-fish etc. have been evolved and popularized in India (Ayyappan *et al.*, 1998; Majhi, 2005). In North Eastern Region of India, integrated fish-livestock farming is a prime activity in majority of the rural households and in recent times more numbers of farmers have adopted the integrated farming practice leaving aside the traditional practice for the reasons; maximum returns with minimum investment and low risk. Thus, the present finding will go a long way in providing basic information about the qualitative and quantitative plankton production by using the livestock manures. In the current study, it has been observed that duck manures perform better over the poultry manure and pig manure in terms of phytoplankton and zooplankton production. Nevertheless, the similar finding was also reported by Olah *et al.*, 1986; Tsirtsis and Karydis, 1998 and

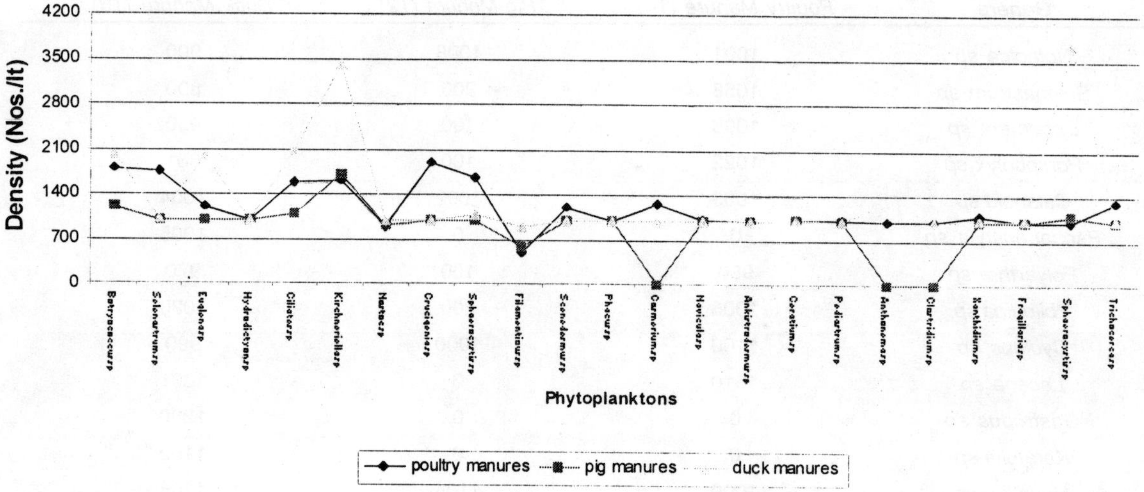

Figure 4.1: Performance of Individual Phytoplankton in Three Treatment Ponds (T1, T2 and T3)

Note: ➡ Showing the quantitatively dominant genera and

◯ indicates the genera that were absent

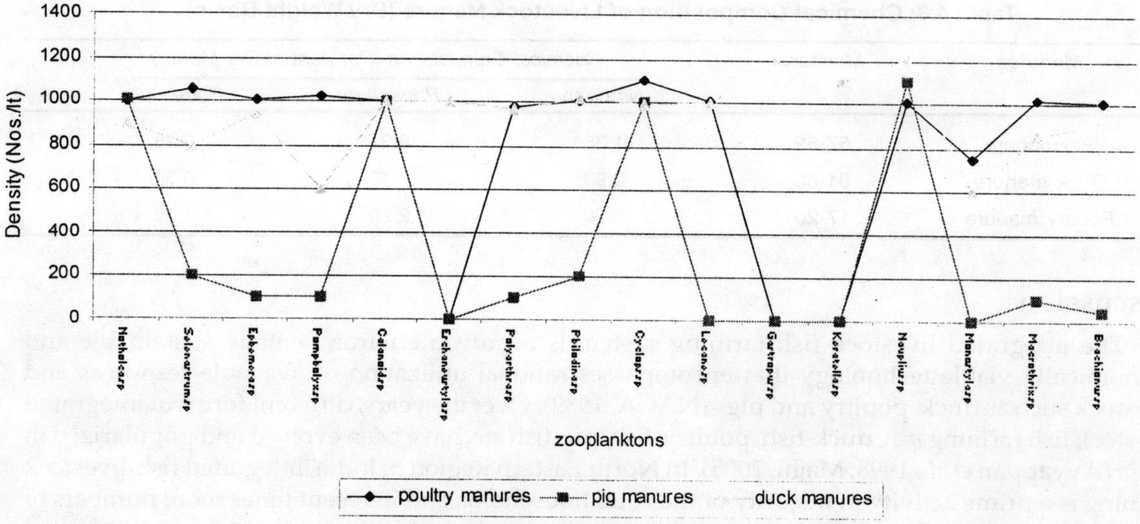

Figure 4.2: Performance of Individual Zooplankton in Three Treatment Ponds (T1, T2 and T3)

Note: ➡ Showing the quantitatively dominant genera and

◯ indicates the genera that were absent

Mukharjee *et al.*, 2001. In our previous work on comparative studies on qualitative and quantitative plankton production using livestock excreta of poultry, pig, duck and goat, we had a similar finding as well (Majhi *et al.*, 2004). Although, proximate analysis revealed that poultry manure to have containing higher percentage of N, P and K over the duck manure and pig manure, but the performance of duck manure was find better in terms of plankton production. This may be attributed by the fact that, every day ducks were allowed to swim in the pond. And, ducks being the bio-aerator might have helped in enriching the water with desired level of oxygen (7.5mg/l) and subsequently the oxygen helping in rapid decomposition of organic matters to release the micronutrients, which results in production of planktons. Nevertheless, further investigation is required to validate the hypothesis.

The biological productivity of any aquatic body is generally judge by the healthy water quality (Ahmed and Singh, 1998) and there exist a direct relation with the fish health and live feed production. In the present study the water quality parameters were found to be within the acceptable range. Sharma and Das (1998) opined that even heavy organic load through livestock manures does not reduce the dissolved oxygen content of the water. Further, Dhawan and Kaur (2002) also reported the similar findings where they had studied the water quality in relation to manuring the pond using pig excreta. However, in recent years, the climatic condition of North Easter Region of India has been in constant change due to global warming, by virtue there is a steady increase in day temperature. Thus, the increasing UV rays intensity coupled with high temperature that reaches up to the pond bottom helps in decomposing the accumulated manures, leaving no risk of obnoxious gas production, which are known to be the poisonous gas responsible for mass mortality of fish biomass. Therefore, slight increase in organic load in the fish pond will not affect the water quality parameters adversely. This is proved in the present study, where every day the pond were applied with livestock manures at the rate of 10 per cent kg body weight of fish biomass.

In recent years, attention has been shifted to develop a mass culture technique of live feeds by using livestock excreta from poultry, duck, pig, cow and goat, as live feeds containing balanced amino acids are the indispensable requisite for growth and gonad maturity of the fish. The North Eastern Hill Region of India has been identified as the potential zone for production of organic foods including fish due to thin population density (151 persons/sq.km) and negligible use of inorganic inputs like fertilizers, growth promoters etc. Therefore, the present information on qualitative and quantitative plankton production using livestock manures would be useful for the rural fish farmers who take up the integrated fish farming as well as the organic farming of high valued food fishes.

Acknowledgements

The author is thankful to the Director, ICAR Research Complex for NEH Region, Barapani for providing necessary facilities to carry out the work.

References

Ahmed, S.H. and Singh, A.K., 1989. Correlation between a biotic factor of water and zooplankton communities of a tank in Patna, Bihar, p. 119–121. In: *Proc. Nat. Sem. on Forty Years of Aquaculture in India*, 7-9 November, Central Institute of Freshwater Aquaculture, Bhubaneshwar, Orissa.

APHA, 1989. *Standard Methods for the Examination of Water and Wastewater*, 17th Edn. American Public Health Association, Washington, DC.

Ayyappan, S., Kumar, K. and Jena, J.K., 1998. Integrated fish farming practices and potentialities. *Fishing Chimes*, 18(1): 15–19.

Dhawan, A. and Kaur, S., 2002. Pig dung as pond manure: Effect on water quality, pond productivity and growth of carps in polyculture system. *NAGA: The ICLARM Quarterly*, 25: 1.

Majhi, S.K., Mahapatra, B.K., Vinod, K. and Mandal, B.K., 2004. Comparative studies on qualitative and quantitative plankton production using different livestock excreta. *Aquacult.*, 5(2): 123–131.

Majhi, S.K., 2005. Prospect of Integrated fish-livestock aquaculture in North Eastern Hill Region of India. *Eco. Env. and Cons.*, 11(2): 287–291.

Mukharjee, T.K., Geeta, S., Rohani, R. and Phang, S.M., 1989. *Proc. Ann. Meeting Mal. Soc. Anim. Prod.*, pp. 4.

NACA, 1989. Integrated fish farming in China. *NACA Technical Manual*, No. 7.

Olah, J., Sinha, V.R.P., Yoppan, S., Purushottam, S. and Radheyshym, S., 1986. Prospectus of Integrated fish farming. *Aquaculture*, 58: 111–122.

Sharma, B.K. and Das, M.K., 1988. Studies on integrated fish-livestock carp farming system. *Fishing Chimes*, 7: 15–27.

Tsirtsis, G. and Karydis, M., 1998. Evaluation of phytoplankton community indices for detecting eutrophic trends in the marine environment. *Environmental Monitoring and Assessment*, 50: 255–269.

Chapter 5

Studies on the Ecology and Fish Fauna of Makroda Reservoir of Guna District (Madhya Pradesh)

Dushyant Kumar Sharma

Introduction

Guna is situated towards northwestern corner of Malwa in southern Gwalior division of Madhya Pradesh. It has several water bodies. The Makroda reservoir is situated at a distance of 15 km from Guna City across the river Negri, about 16 km upstream of Rampur tank near village Bhairoghati. Negri river is a tributary of Parwati river of Chambal basin. The reservoir has a catchment area of approximately 65.16 sq miles. The reservoir serves the purpose of irrigation. The water is utilized for irrigation of 26055 acres of land in 32 villages of Bamori block (Guna). It is also used for pisciculture of major and minor carps and locally available fishes. The present work was undertaken to investigate the physico-chemical parameters of the reservoir and their effect upon fish life.

Materials and Methods

Monthly periodical studies were carried out for the year October 2003 to September 2004. The water and fish samples were collected between 8-11 A.M. Standard methods by APHA (1981) and Trivedy and Goel (1986) were followed for physical and chemical analysis of the water. Fish samples were collected and were identified using Day (1989), Jhingran (1975), Qureshi (1983) and Srivastava (1986).

Results and Discussion

Variations in physico-chemical parameters are presented in Table 5.1. The fishes are enlisted in Table 5.2.

Table 5.1: Physico-chemical Analysis of Makroda Reservoir, Madhya Prdesh

Months	*Temperature (°C)*	*Trans-parency (cm)*	*pH*	*Free Carbon Dioxide (ppm)*	*Dissolved Oxygen (ppm)*	*Alkalinity (ppm)*	*Hardness (ppm)*	*Chlorides (ppm)*
Oct-03	26.2	23.6	7.6	5.7	4.9	82	33.5	6.1
Nov-03	21.2	39.5	7.8	5.6	5	94	33	6
Dec-03	20	40.2	7.9	5.4	5.6	98	31.2	5.4
Jan-04	17.1	41.2	8.1	4.1	6.	95	31.3	5.5
Feb-04	18.4	3237	8.1	4.2	6.4	93	34.5	5.8
Mar-04	21.6	36.4	7.9	3.9	6.5	99	34.2	6
Apr-04	24	35.6	8	4.1	8.1	101	38.1	6.1
May-04	28.9	28.3	8.2	3.6	8.9	111	42.5	6.5
Jun-04	30.4	20.1	8.5	2.5	9	114	56	8.2
Jul-04	29.8	14.5	8.5	6.5	4.4	91	55	8.5
Aug-04	28.2	13.2	8.1	8	4.6	80	47	7.1
Sep-04	27	20.5	7.9	10.4	4.8	84	45	6.9

Table 5.2: List of Fishes Recorded from Makroda Reservoir, Madhya Pradesh

Sl.No.	*Scientific Name of the Fish*	*Local Name*	*Family*	*Fin Formula*
1.	*Labeo rohita*	Rohu	Cyprinidae	D.16(3/13);P.17;V.9;A.7(2/5);C.19;L.I.40-41;L.tr.61/2-71/2/9 Barbels 1 Pair
2.	*Puntius ticto*	Sidhari	Cyprinidae	D.11(3/8);P.13;V.9;A.8(3/5);C.19;L.I.25;L.tr.51/2-61/2
3.	*Catla catla*	Katla	Cyprinidae	D.17-19(3/15/16);P.20;V.8;A.8(3/5);C.18-20;L.I.43;L.tr.71/2-61/2
4.	*Cirrhinus mrigala*	Nain	Cyprinidae	D.16(3/13);P.18;V.9;A.8(2/6);C.15;L.I.42;L.tr.61/2-71/2 Barbels 1 Pair
5.	*Wallago attu*	Parhin	Siluridae	D.5;P.1/14;V.10;A.86(4/82);C.17 Barbels 2Pairs

The results show that the water temperature varies from 17.1°C to 30.4°C with the atmospheric temperature *i.e.* being maximum during summer (30.4°C in June 2004) and minimum during winter (17.1°C) the fluctuations in water temperature has relationship with air temperature. A direct relationship between air and water temperature was also found by Singhai *et al.* (1990), Sharma and Jain (2000), Khanna and Singh (2000) and Khanna *et al.* (2003). The water temperature was found well within the limits of survival of fishes.

Transparency was also quite favourable for fish fauna. Maximum transparency was found during winter and minimum during rainy season. The minimum transparency during rainy season was due to high turbidity of water caused by suspended slit and organic debris (Sahai and Sinha 1969, Balkhi *et al.*, 1987,Dagaonkar and Saxena 1992 and Khanna 1993) Turbidity of the water may be either due to phytoplankton or due to suspended particles. During summer, low values of transparency may be attributed to the growth of phytoplankton. Jana (1979) and Bhatt and Negi (1985) also reported similar findings.

During the study period pH variation lies mostly within the alkaline range. According to Jhingran (1977) pH range from 7 to 9 is considered good for fish culture. In the present study, the observations show the water to be suitable for fish culture.

The dissolved oxygen ranged from 4.4 to 9 ppm having moderate fluctuations. Free carbon dioxide varies considerably throughout the year. The presence of free CO_2 is due to incomplete utilization in photosynthesis. Free CO_2 and dissolved oxygen showed inverse relationship to one another similar to observations of Welch (1952),Hutchinson (1957) and Hynes (1970). The values were in the range of fish life.

Alkalinity of the water is characterized by the presence of all hydroxyl ions capable of combing with the hydrogen ions. Total alkalinity shown in the present study shows seasonal variations. The values were high during summer. Goldman and Wetzel (1963) and Sharma (1992) also reported similar results.

The chloride concentration exhibited seasonal variations. It was high during summer and low during winter. The high values during summer may be attributed to low water levels (Gonzalves and Joshi, 1946). According to Munawar (1970) the higher values of chloride in the water are an index of pollution of animal origin. The values of chloride were not found to be very high, throughout the year, in the present study. Chloride and hardness show positive relationship. Similar type of relationship was also observed by Khanna *et al.* (1999).The values were quite favourable for fish life.

Summary

Makroda reservoir is the one of the important artificial water bodies of Guna. The water is used by the nearby villagers mainly for irrigation and pisciculture. Physico-chemical studies were undertaken to enhance the limnological knowledge about the water body and to explore the possibilities for better management of pisciculture. The study was undertaken for the period of October 2003 to September 2004. Considerable seasonal variations were noted in various physico-chemical factors.Altogether the total number of fishes described are ten in number. Study of various physico-chemical factors also indicated that the water is quiet suitable for pisciculture and by proper methods water can be properly utilized and exploited.

References

APHA-AWWA-WPCF, 1981. *Standard Methods for the Examinations of Water and Wastewater*, 15th ed. APHA, Washington D.C.

Balkhi, M.H., 1987. Hydrobiology of Anchor lake Kashmir. *Physiol. Eco.*, 12(3): 131–139.

Bhatt, S.D. and Usha Negi, 1985. Physico-chemical features of phytoplankton population in a subtropical pond. *Comp. Physiol. Ecol.*, 10(2): 85–88.

Dagaonkar, A. and D.N. Saxena, 1992. Physico-chemical and biological characterization of a temple Tank, Kaila Sagar, Gwalior, M.P. *J. Hydrobiol.*, 8(1): 1–19.

Day, F., 1989. *Fauna of British India, including Cyclone and Burma Fishes* I, 548 pp. London Taylor and Francis.

Goldman,C.R. and R.G. Wezel, 1963. A study of primary productivity of clear lake country calibormis. *Ecology*, 44: 285–294.

Gonzalves, E.A. and D.B. Joshi, 1946. Fresh water algae near Bombay: The seasonal succession of the algae in a tank of Bandra. *J. Bombay Nat. Mist. Soc.*, 46: 154–176.

Hutchinson, G.E., 1975. *A Treatise on Limnology*, Vol. 1. John Wiley and Sons, New York, London, 1015 pp.

Hynes, H.B.N., 1970. *The Ecology of Running Waters*. Liverpool University Press, 4th impression.

Jana, B.B., 1979. Temporal plankton succession and ecology of a tropical Tank in West Bengal, India. *Int. Rev. Gas Hydrobiol.*, 64(5): 661–671.

Jhingran,V.G., 1975. *Fish and Fisheries of India*, 1st ed. Hindustan Publishing Co., Delhi, India.

Khanna, D.R. and R.K. Singh, 2000. Seasonal fluctuations in the Plankton of Suswa River at Raiwala (Dehradun). *Environment Conservation Journal*, 1(2 and 3): 89–92.

Khanna,D.R., 1993. *Ecology and Pollution of Ganga River*. Ashish Publishing House, New Delhi.

Khanna, D.R., Shakun Singh, Ashutosh Gautam and J.P. Singh, 2003. Assessment of water quality of river Ganga in District Bulandshahar (U.P.) India. *J. Natcon.*, 15(1): 167–175.

Khanna, D.R., T.R. Seth, D.S. Malik and R.K. Singh, 1999. Fishes and the ecology of the river Suswa at Raiwala Dehradun. In: *Sustainable Ecosystems and Environment*, (Eds.) D.R. Khanna, Ashutosh Gautam and Aditya Goutam, p. 105–110.

Munawar, M., 1970. Limnological studies of fresh water ponds of Hyderabad, India. *Hydrobiologia*, 35: 127–162.

Qureshi, T.A., 1983. *Indian Fishes*. Brej Brothers, Bhopal.

Sahai, R. and A.B. Sinha, 1969. Investigations on Bioecology of inland of Gorakhpur (U.P.) India I Limnology of Ramgarh lake. *Hydrobiol.*, 34(3–40): 433–447.

Sharma, M., 1992. Hydro biological studies of Halali Reservoir with reference to zooplanktons and fishery prospects. *Ph.D. Thesis*, Baraktulla University, Bhopal M.P.

Sharma, Dushyant and Renu Jain, 2000. Physico-chemical analysis of Gopalpura tank of Guna District (M.P.). *Ecol. Env. and Cons.*, 6(4): 441–445.

Shrivastva, G.L., 1986. *Fishes of Eastern U.P.* Vishwavidyalaya Prakashan,Varanasi.

Singhai, S., G.M.A. Ramani and U.S. Gupta, 1990. Seasonal variations and relationship of different physico-chemical characteristics in newly made Tawa reservoir. *Limnologica* (Berlin), 21(1): 263–301.

Trivedy, R.K. and P.K. Goel, 1984. *Chemical and Biological Methods for Water Pollution Studies*. Environmental Publications, India.

Chapter 6

Marine Finfish Resources of Nagapattinam Coast, Tamil Nadu, South India

M. Rajasegar, J. Gopal Samy and B. Bama Devi

Introduction

Marine and coastal biodiversity encompasses the enormous variety of marine species and their biological diversity is always high. India is being one of the mega diversity country has vast coastal line of 8500 kms. But information on biodiversity of different groups are lacking in different locations. Mostly of the data are available only on selected known sites. Though there are severe damage has been posed on the biodiversity still we are unclear that what natural biological resources these coastal reserves possess and what we are losing at present because of lack of data on the biodiversity of these regions. Hence the present study has been made to determine the finfish resources occurring along the Nagapattinam coast and to document the biodiversity of the finfishes of this coast, to assess the current status of knowledge and critically elaborating the current status such endanger, endemic, abundant, rare, economically and socio-culturally important species or groups.

Materials and Methods

Nagapattinam coast is situated in Coromandel coast of India.There are four stations selected from this coast *i.e.* Point Calimere, Nagapattinam, Tharangambadi and Pazhayar. Specimens of finfishes were collected from landing centres and markets during July 2005 to June 2006 and preserved in 10 per cent formalin. Later specimens were examined the various morphometric characters for identification and the works of Munro (1955), Fischer and Bianchi (1984), Talwar and Kacker (1984) and De Bruin *et al* (1994) were referred for confirmation of identification. The study period was divided into four seasons *viz.*, premonsoon (July to September), monsoon (October to December), post monsoon (January to March) and summer (April to June) based on the north east monsoon.

Results and Discussion

In the present study, 169 species of finfish were recorded from the Nagapattinam coast. Among these, *Stolephorus indicus, Thryssa malabarica, Chirocentrus dorab, Arius arius, Mugil cephalus, Platycephalus indicus, Epinephelus tauvina, E. malabarica, Sillago sihama, Caranx sem, Gerres abbreviatus, Upeneus sulphereus, Drepane punctata, Terapon puta, Cynoglossus lingua* and *Triacanthus biaculetus* were abundant throughout the study period at all the four stations. *Pampus argenteus, P. chinensis, Trichiurus leptutus, Ephinephelus coioides, Leignathus daura, Lutjanus gibbus, L. lineolatus, Trachinocephalus myops, Carangoides malabaricus* and *Aesopia cornuta* were abundant only in premonsoon at Nagapattinam and Pazhayar stations only.

Himantura uarnak, Liza vaigiensis and *Mene maculata* were abundant during monsoon season and common during postmonsoon at all the stations. Endangered species like *Carcharhinus limbatus* and *C. palasurah* were recorded during all the seasons in few numbers from all the four stations. Likewise another globally endangered genera *Hippocampus* represented by *H. kuda* was recorded at Nagapattinam and Point Calimere in very few number. *Rachycentron canadum* was recorded rarely at Nagapattinam alone.

The seasonal fishes such as *Cypelurus spilopterus, Hirundichthys coromandelensis* and *Exocoetus volitans* were abundant only during April to July and they are inconspicuous in rest of the year. Their distribution is also differed spatially as they have been recorded only from Nagapattinam and Tharangambadi stations alone. The previous record showed (Rajasegar, 2004), these species were recorded during middle of May to July. But during the present study, they were recorded middle of April *i.e.* one month before the actual season. Another interesting aspect is discontinuous distribution of *Harpodon nehereus* which is abundantly available in Nagapattinam station. Previous records showed their occasional landing in few numbers from this station. This change in species composition and population dynamics might be due to tsunami.

Though there are numerable studies on finfish diversity their spatial and temporal variations are available throughout the Indian coast (Whitehead,1972;Kalyan Chakrabarti, 1976; Talwar and Kacker, 1984; Shameem and Dutt, 1986; Ramaiyan *et al.*, 1987; Badal Chandra Bharati Goswami, 1992; Sujatha, 1995; Ramaiyan and Senthilkumar, 1997; George, 1998; Ramaiyan *et al*, 2002.). Such studies are lacking in this part of the coast though these stations are having long history of fishing activity and tradition. Rajasegar (2004) is the only available report on finfishes from the Karaikal coast. Under the light of the above conditions, the present investigation on the finfish diversity of coastal and marine waters along the Nagapattinam coast has brought out two important aspects. One is the availability of *Harpodon nehereus* in large quantities and the second is the change of season. *i.e.* the occurrence of flying fish one month before the actual period. Since the trash fishes are left in the sea itself the present results would not brought out the original pictures of finfish diversity of this coast. If that also be estimated with seasonal differences there are chances of identification of new edible fishes as there is continuous reduction in standing stock of the most of all edible fishes. This would supplement and cater the increasing fish market.

Conclusion

Marine finfish diversity was studied from Nagapattinam coast during July 2005 to June 2006. During the present study, 169 species of finfish belonging to 15 orders, 68 families and 117 genera were recorded from the four stations. Among these, *Stolephorus indicus, Thryssa malabarica, Chirocentrus dorab, Arius arius, Mugil cephalus, Platycephalus indicus, Epinephelus tauvina, E. malabarica, Sillago sihama, Caranx sem, Gerres abbreviatus, Upeneus sulphereus, Drepane punctata, Terapon puta, Cynoglossus lingua* and *Triacanthus biaculetus* were abundant throughout the study period at all the four stations. Seasonal abundance, seasonal variations in species composition were also observed. The present investigation on the finfish diversity of coastal and marine waters along the Nagapattinam coast has brought out two important aspects. One is the availability of *Harpodon nehereus* in large quantities and the second is the change of season, *i.e.* the occurrence of flying fish one month before the actual period. Since the trash fishes are left in the sea itself the present results would not brought out the original pictures of finfish diversity of this coast.

Acknowledgement

The authors thank to the Principal and Head of the Department of Zoology, TBML College for their support and facilities.

References

Badal, Chandra and Bharati Goswami, 1992. Marine fauna of Digha coast of West Bengal, India. *J. Mar. Biol. Asso., India*, 34: 115–137.

De Bruin, G.H.P., B.C. Russel and A. Bogusch, 1994. *The Marine Fishery Resources of Sri Lanka*. FAO species identification field guide for fishery purpose. Rome FAO, 400 p.

Fischer, W. and G. Bianchi, 1984. *FAO Species Identification Sheets for Fishery Purposes*. Western Indian Ovean (Fishery Area 51). FAO, Rome, Vols. I–IV: Pag.Var.

George, V. I., 1998. *Commercial Fish and Shell Fishes of India*, CMFRI, 177 pp.

Kalyan, Chakrabarti, 1976. *Mahasagar. Bull. Natl. Ins. Oceanogr.*, 19: 147–151.

Munro, I.S.R., 1955. *The Marine and Freshwater Fishes of Ceylon*. Dept. of External Affairs, Camberra, 351 pp.

Rajasegar, M., 2004. Coastal and marine finfish resource occurring along Karaikkal coast. *M. Sc. Dissertation*, Sikkim Manipal University, India, 30 pp.

Ramaiyan, V., Purusothaman, A. and Natarajan, R., 1987. Checklist of estuarine and marine fishes of Parangipettai (Porto Novo) coastal waters. *Matsya*, 12–13: 1–19.

Ramaiyan,V., R. Senthilkumar and M. Rajasegar, 2002. *Finfish Resources of Pitchavaram Mangroves Ecosystem, Monograph*. Annamalai University, India, 94 pp.

Ramaiyan, V. and R. Senthilkumar, 1997. *Laboratory Manual on Fisheries Science*. Annamalai University, India, 33 pp.

Shameem, A. and Dutt, S., 1986. Fishes of the family caranagidae from Visakhapatnam: List of species and new records from the Indian seas. *J. Mar. Biol. Asso. India*, 28: 159–162.

Sujatha, K., 1995. Finfish constituents of trawl by catch off Visakhapatnam. *Fishery Technology*, 32: 56–60.

Talwar, P.K. and Kacker, R.K., 1984. Commercial sea fishes of India. *Rec. Zool. Sur. India*, 997 pp.

White Head, P.J.P., 1972. A synopsis of the clupeoid fishes of India. *J. Mar. Biol. Ass., India*, 14 :160–256.

Chapter 7

Microbial Ecology of Groundwater at Lucknow City

D.K. Singh, P.K. Bharti and D.S. Malik

Introduction

Water is essential to sustain life. Safe drinking water is one of the prime necessities of human health. Ever increasing urbanization has caused microbial contamination of various drinking water sources. A wide variety of viruses, bacteria, protozoa, fungi, algae and helminths are the major groups of pathogens of common occurrence in contaminated water, impart a great deal of water borne diseases. Sporadic incidences as well as epidemics of various water borne diseases like gastroenteritis, cholera, dysentery, typhoid, etc. are the common health problems of global concern. As per World Health Organization (WHO), standard and guidelines for drinking water (WHO, 1993), it should be free from faecal organisms, pathogens and toxicants because the contaminated water results in mortality of millions of world population.

The problem of water-transmitted diseases has commonly been considered to be associated with aquatic pathogens from faecal wastes and municipal sewage. The presence of these pathogens in water can be assessed by determining the density of a group of certain members belonging to the family Enterobacteriacae. Some commonly occurring coliforms are *Escherichia coli, Klebsiella sp., Enterobacter sp., Citrobacter sp.* and *Streptococcus sp.* etc. The coliform group of bacteria has universally been used for more than 80 years as prime indicator to assess the biological contamination of water. The coliform group is currently defined as aerobic, Gram-negative rods, which ferment lactose with gas production within 48 hours at 35°C (APHA, 1995).

The coliform organisms are ubiquitous in nature. It is well known that non-faecal coliform bacteria can grow in water even with very small amounts of organics. Coliform can survive for several months in natural tropical river water. The Indian Standard (IS: 10500-1991) for drinking water recommends an average MPN of coliform not exceeding 10 organism per 100 ml with no faecal coliform (BIS, 1996).

Coliform bacteria are the natural part of the microbial flora of the intestinal tract of warm-blooded mammals, including man. Coliform bacteria can also be found in soil, vegetation and other animals. The coliform group is relatively easy to culture in the lab and therefore, has been selected as primary indicator bacteria for the presence of disease causing organism. Coliform bacteria are normally non-pathogenic organisms and are only mildly infectious. For this reason these bacteria are relatively safe to work with in the laboratory. If large number of coliform is found in water, there is a high probability of presence of other pathogenic viruses, bacteria, protozoa and helminthes.

For more than 100 years, bacteriological tests have been used to detect faecal contamination of drinking water. During this time, there has been an evolution in the bacterial indicators used and the articulation of the criteria for an ideal or reliable indicator of faecal contamination in drinking water. The current criteria of an ideal or preferred indicator of faecal contamination have been defined and stated by WHO and other authorities. According to these authorities the essential criteria of a faecal indicator, are the following (WHO):

1. The indicator should be absent in unpolluted water and present when the source of pathogenic microorganisms of concern (faecal contamination) is present.
2. The indicator should be present in greater number than the pathogenic microorganisms.
3. The indicator should respond to natural environmental conditions and water treatment processes in a manner similar to the pathogens of concern.
4. The indicator should be easy to isolate, identify and enumerate.
5. The tests should be inexpensive thereby permitting numerous analyses to be made.
6. The indicator should not be a pathogenic microorganism.
7. The indicator should not multiply in the environment.

Therefore, any indicator of faecal contamination of drinking water and its sources is best judged according to these essential criteria and perhaps other criteria that may be relevant for or specific to a particular indicator and type of tests for it.

E. coli is normal inhabitant of the intestine, and most strains are non-pathogenic. Isolation of *E. coli* from a water supply indicates faecal contamination. Members of lactose fermenting species of this group, which may be referred to colloquially as "coliform" occur in a variety of niches not all of which are intestinal. The picture is further complicated by the fact that other members of the coliform group are also found in the intestine. Thus the definitive identification of *E. coli* may be needed to determine the significance of "coliforms" in a water supply (WHO, 1993). *Klebsiella* is a thermotolerant coliform indicator of contaminated water. The genus *Klebsiella* consists of non-motile, capsulated rods that grow well on MacConkey agar media. The total coliform bacterial test is primary indicator of 'potability' (suitability for consumption) of water sample. It determines the level of total coliform bacteria associated with the possible presence of disease causing organisms.

E. coli are used as a prime indicator of faecal coliform bacteria. The genus is named after Escherich who was the first to describe the colon bacillus, under the name *Bacterium coli commune* in 1885. MacConkey had prepared a medium in 1905 by means of which the presence of *Bacterium coli* could be ascertained more easily and certainly other test receded into the background but maintained their importance as alternatives in the detection of faecal pollution. *E. coli* is an aerobe and a facultative anaerobe. The temperature range is 10-40°C (optimum 37°C). It grows well on MacConkey agar medium. Colonies are thick, bright pink coloured due to lactose fermentation. In MacConkey broth, it changes the colour of broth from red to yellow due to the acid production by lactose fermentation. Glucose, Lactose, Mannitoal, Maltose and many other sugars are fermented with the production of acid and gas. The four-biochemical tests widely employed in the classification of enterobacteria are the indole, methyl-red, Voges-Proskauee (VP), and citrate utilization tests, generally referred to by "IMViC" test.

Faecal streptococci are other indicator bacteria. Their presence indicates that the water is contaminated. In the genus *Streptococcus*, only *S. bovis* and *S. equines* are included in the faecal streptococcus group. Their sources are mainly animal faeces. Streptococci rarely multiply in polluted water and they are more resistant than *E. coli* and other coliform bacteria. Their primary value in water quality examination is therefore as additional indicator of treatment efficiency (WHO, 1999).

Streptococci ferment several sugars, producing acid but no gas. Streptococci are catalase negative. Along with humans, other sources of Streptococci are excreta from various domestic and wild animals, birds, reptiles, fishes and insects etc. These organisms have been considered to be an index of faecal pollution of drinking water. The ratio of faecal streptococci and thermotolerant coliforms may be used to identify the sources of pollution *i.e.* excreta from human beings or from homoeothermic animals. This is due to the presence of more coliform than *Streptococci* in human faeces.

Emergence of antibiotic resistance is directly associated with indiscriminate and excess use of antibiotics for treatment of infectious diseases. In our country most of the natural resources of water are contaminated with coliforms exhibiting increased resistance to various antibiotics (Ramteke *et al.*, 1991). A significant impact on the incidence of resistance has been reported in India after epidemics of gastroenteritis by causal agent was identified as multiple antibiotic resistant organisms. The antibiotic resistance in bacteria has often been found in many cases to be plasmid borne. It is known that plasmid, is a self-replicating extra-chromosomal genetic material. Antibiotic resistance in organisms that are not primary pathogens is important because of their ability to transfer their resistant plasmid (R-plasmid) to sensitive pathogenic organisms resulting in epidemics of antibiotic resistant pathogens that may pose a great problem in effective treatment.

Drug resistance in bacteria may be brought about by the production of drug destroying enzyme. This is a situation seen in the destruction of Penicillin by penicillinase enzyme. This mechanism accounts for only a minority of drug-resistant organisms; most become tolerant to the antibiotics. The mechanisms by which, organisms become antibiotic tolerant vary with different antibiotics, but usually involve the use of an antibiotic. The mechanisms by which antibiotic-resistant organisms become predominant in a community have often been considered to by mutation to antibiotic resistance.

As far as possible, water sources must be protected from contamination by human and animal waste, which can contain a variety of pathogens and parasites. Failure to provide adequate protection and effective treatment will expose the community to the risk of outbreaks of intestinal and other infectious disease. The literature survey has revealed that a lot of work has been done in the field of aquatic microbiology in relation to bacterial population in drinking water. In view of these, an attempt has been made to enumerate, isolate, identify and characterize the microbial population in the drinking water of Lucknow city. Microorganisms isolated from the drinking water were studied in relation to pathogenicity and resistance-phenomenon.

Materials and Methods

The samples were collected in a pre-sterilized, heat-resistant, borosilicate glass, stopper bottle of required capacity. Sampling was done aseptically, directly into the bottle from the sources to avoid contamination. The samples were kept cool (4–10°C) during transport to the laboratory and inoculated within 6 hours of collection, for analysis by the multiple tube fermentation method. In this method, a series of dilutions with 10ml, 1ml, and 0.1ml volume of sample in three replicates were inoculated. Production of gas with/without acid production within 24 hours at 35°C was main criterion for positive results. A statistical analysis is used to determine the most probable number (MPN) of bacterial cells per 100ml. of sample.

The water samples have been collected from different residential areas of Lucknow city, during April–May 2004. Lucknow is capital of Uttar Pradesh and has its own historical, archeological, cultural and academic identity. The samples for different microbial parameter were collected from Aliganj,Vikas Nagar, Indira Nagar, Gomti Nagar, Paper Mill Colony, Nirala Nagar, Triveni Nagar, Rajajipuram, Model House, Ashiana, Sadar Cantt., Charbagh, Alambagh, Chowk, Hazratganj, Hussainganj, Aminabad, Amausi, Talkatora and Aishbagh sampling sites, and analyzed with the help of the procedure described by APHA (1995), Trivedi and Goel (1984).

Results and Discussion

The ever-growing urbanization and industrialization are continuously polluting the water bodies,containing sewage pollutant and other types of waste material drinking water is a prime necessity for life and by uncontrolled civilization it become more polluted. Drinking waters are exposed to microbes and toxicants in water. As consumer of water *i.e.* human beings are also indirectly exposed to that water contamination.

In this study, the bacterial density as MPN/100 ml. was observed. The hundred samples were collected and tested for coliforms, faecal coliforms and faecal streptococci. Coiforms/100ml were observed to be <3–>100 MPN/100ml. and faecal coliform/100ml were observed to be <3–>1100 MPN/100ml. where as faecal streptococci were also observed to be <3–460 MPN/100ml. We were divided the sampling sites into three categories *i.e.* residential, commercial and industrial areas. In these areas the maximum faecal pollution was found in the commerçial areas, about 27 per cent and in residential and industrial areas have 24 per cent faecal pollution. These observations indicate that

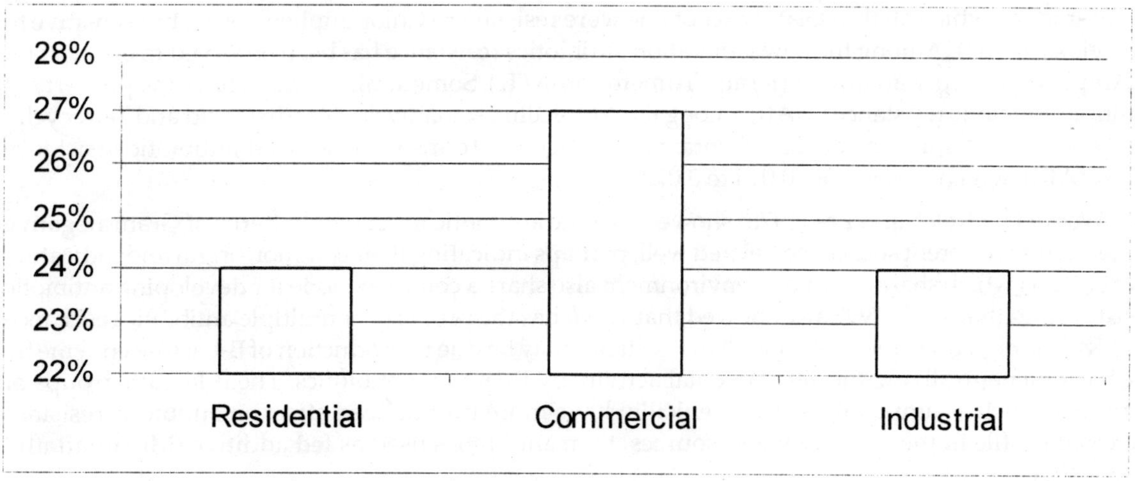

Figure 7.1: Contaminated Percentage of Different Areas

faecal pollution is maximum in those areas. Where proper sanitation is not handled, such as Charbagh, Aminabad etc.

The random isolation, purification and biochemical characterization of 34 bacterial strains from the water samples of different areas showed a wide range of bacterial species. Only few strains are belonging to the category of pathogenic bacteria, otherwise most of them were probably non-pathogenic or opportunistic pathogens.

The coliform were identified on the basis of biochemical tests and their Gram-staining. Coliforms are mainly rod-shaped and Gram-negative most of them ferment the sugars and produce gas. American public health association (APHA, 1995) defined the coliform group as including "all of the aerobic and facultative anaerobic, Gram-negative, non-spore forming, rod-shaped bacteria which ferment lactose with gas formation and also produce acid within 48 hours at 35°C.

Identification of bacterial strains was observed on the basis of their biochemical characterization. The IMViC test is most sufficient in identification but catalase test, urease test, motility test; H_2S production, oxidase test and Gelatin liquifection test were also permored.

As mentioned earlier that most of the water sources are being contaminated with municipal sewage containing faecal coliform bacteria. Since in these days a wide range of antibiotics are used in treatment of different infections diseases. The gut flora of a patient after long exposure to such therapeutic chemicals, a vast bacterial population may affect the plasmid in bacterial cells.

Transmissibility of such plasmid-mediated resistance of bacteria to nay other sensitive bacteria is a significant especially of this characteristic. In this way the population of resistant bacteria increases and this phenomenon has posed a therapeutic problem in treatment of infectious disease in human being. Antibiotic resistant were observed by disc diffusion method. Doses of ten antibiotic (g) *viz.*, Amoxyellin (10), Chloramphenicol (30), Ciprofloxacin (100), Kanamycin (30), Gentamycin (10), Nalidixic acid (30), Norfloxacin (10), Streptomycin (10), Tetracyclin (30) and Trimethoprim (05) were use in this study.

Earlier studies in this field have revealed that environmental and clinical isolates of bacteria have developed resistance for many of the new generation antibiotics. This fact is evident from the findings of my study in which all the isolates except one were resistant to Chloramphenicol (C) but sensitive to ciprofloxacin (Cf). Among the new generation antibiotics resistance has been observed to be resistant to Amplicillin (Am), Kanamycin (K) and Trimethoprim (Tr). Some strains of *E.coli* have the property of multiple antibiotic resistance (MAR) among the Ampicillin, Kanamycin, Nalidixic acid and Tetracyclin but sensitive to Ciprofloxacin and Gentamycin. The test strains were showed antibiotic resistance index (ARI), which varies from 0.014 to 0.023.

Worked by Kelch and Lee (1978) showed that the antibiotic resistance patterns of Gram-negative bacteria from different sources correlated well, perhaps indicating their common origin and indicating that bacteria which share a common environment also share a common mode for developing antibiotic resistance. Pathak *et al.*, 1993 also showed that *E. coli* has the property of multiple antibiotic resistance (MAR). The high occurrence of ampicillin resistance may be due to production of B-Lactose enzyme by the bacteria neutralizing the antimicrobial activity by B-lactum antibiotics. The B-lactum groups of antibiotics such as ampicillin and carbenicillin have more pronounced effect on antibiotic resistant bacterial profile in the primary water sources, than antibiotics used as fed additive (Mulamattathil *et al.*, 2000).

Table 7.1: Most Probable Number (MPN) Table Used for Evaluation of the Data in the Experiments, Using Three Tubes in Each Dilution

Number of Positive Tubes in Dilutions				*Number of Positive Tubes in Dilutions*			
10 ml	*1ml*	*0.1 ml*	*MPN per 100 ml*	*10 ml*	*1 ml*	*0.1 ml*	*MPN per 100 ml*
0	0	0	<3	2	0	0	9
0	1	0	3	2	0	1	14
0	0	2	6	2	0	2	20
0	0	3	9	2	0	3	26
0	1	0	3	2	1	0	15
0	1	1	6	2	1	1	20
0	1	2	9	2	1	2	27
0	1	3	12	2	1	3	34
0	2	0	6	2	2	0	21
0	2	1	9	2	2	1	28
0	2	2	12	2	2	2	35
0	2	3	16	2	2	3	42
0	3	0	9	2	3	0	29
0	3	1	13	2	3	1	36
0	3	2	16	2	3	2	44
0	3	3	19	2	3	3	53
1	0	0	4	3	0	0	23
1	0	1	7	3	0	1	39
1	0	2	11	3	0	2	64
1	0	3	15	3	0	3	95
1	1	0	7	3	1	0	43
1	1	1	11	3	1	1	75
1	1	2	15	3	1	2	120
1	1	3	19	3	1	3	160
1	2	0	11	3	2	0	93
1	2	1	15	3	2	1	150
1	2	2	20	3	2	2	210
1	2	3	24	3	2	3	290
1	3	0	16	3	3	0	240
1	3	1	20	3	3	1	460
1	3	2	24	3	3	2	1100
1	3	3	29	3	3	3	>1100

Table 7.2: Bacteriological Quality of Drinking Water of Lucknow City

Sl.No	Area	No	Coliform/100ml	Faecal Coliform/ 100ml	Faecal Streptococci/100ml
1.	Aliganj	1.	43	<3	<3
		2.	<3	<3	<3
		3.	<3	<3	4
		4.	23	4	<3
		5.	<3	<3	21
2.	Vikas Nagar	6.	150	<3	<3
		7.	<3	<3	<3
		8.	<3	<3	3
		9.	<3	<3	3
		10.	<3	<3	<3
3.	Indira Nagar	11.	<3	<3	9
		12.	15	<3	4
		13.	<3	<3	<3
		14.	<3	<3	7
		15.	<3	<3	<3
4.	Gomti Nagar	16.	<3	<3	<3
		17.	<3	<3	7
		18.	93	4	<3
		19.	<3	<3	4
		20.	<3	<3	<3
5.	Papermill Colony	21.	<3	<3	<3
		22.	<3	<3	<3
		23.	<3	<3	<3
		24.	15	<3	<3
		25.	<3	<3	<3
6.	Hazratganj	26.	1100	15	7
		27.	<3	<3	<3
		28.	<3	<3	<3
		29.	<3	<3	<3
		30.	7	<3	<3
7.	Charbagh	31.	<3	<3	<3
		32.	1100	460	75
		33.	>1100	93	120
		34.	<3	<3	4
		35.	<3	<3	<3
8.	Alambagh	36.	<3	<3	<3
		37.	<3	<3	4
		38.	23	4	<3
		39.	<3	<3	<3
		40.	>1100	>1100	4

Table 7.3: Bacteriological Quality of Drinking Water of Lucknow City

Sl.No	Area	No	Coliform/100ml	Faecal Coliform/ 100ml	Faecal Streptococci/100ml
9.	Amousi Airport	41.	<3	<3	<3
		42.	<3	<3	9
		43.	23	23	20
		44.	<3	<3	7
		45.	<3	<3	4
10.	Talkatora	46.	<3	<3	<3
		47.	<3	<3	4
		48.	<3	<3	<3
		49.	75	20	<3
		50.	<3	<3	<3
11.	Nirala Nagar	51.	93	9	9
		52.	9	<3	<3
		53.	<3	<3	<3
		54.	<3	<3	<3
		55.	75	23	23
12.	Triveninagar	56.	<3	<3	11
		57.	<3	<3	3
		58.	<3	<3	<3
		59.	<3	<3	<3
		60.	<3	<3	<3
13.	Chowk	61.	7	4	4
		62.	<3	<3	4
		63.	4	<3	3
		64.	<3	<3	4
		65.	240	15	15
14.	Rajaji Puram	66.	4	<3	<3
		67.	3	<3	11
		68.	4	<3	6
		69.	<3	<3	<3
		70.	7	<3	<3
15.	Aishbagh	71.	<3	<3	<3
		72.	<3	<3	<3
		73.	>1100	75	3
		74.	23	<3	4
		75.	<3	<3	<3
16.	Aminabad	76.	<3	<3	4
		77.	4	<3	11
		78.	<3	<3	9
		79.	<3	<3	3
		80.	<3	<3	23

Table 7.4: Bacteriological Quality of Drinking Water of Lucknow City

Sl.No	Area	No	Coliform/100ml	Faecal Coliform/ 100ml	Faecal Streptococci/100ml
17.	Model House	81.	<3	<3	<3
		82.	23	9	3
		83.	<3	<3	4
		84.	<3	<3	3
		85.	4	<3	<3
18.	Hussainganj	86.	<3	<3	<3
		87.	4	<3	4
		88.	<3	<3	7
		89.	<3	<3	<3
		90.	<3	<3	290
19.	Sadar Cantt	91.	4	4	460
		92.	<3	<3	9
		93.	<3	<3	<3
		94.	<3	<3	4
		95.	<3	<3	39
20.	Ashiana	96.	<3	<3	7
		97.	7	7	23
		98.	14	9	43
		99.	43	7	43
		100.	<3	<3	<3

Furthermore, about two or three bacterial strains in my study have shown sensitivity for one or two antibiotics as well as other bacterial strains were showed multiple antibiotics resistance other workers have also reported for variable number of antibiotics among of coliform and heterotrophic bacteria in water (Armstrong *et al*, 1981). Multiple antibiotic resistance patterns have been reported among the aquatic bacterial population and coliform bacteria (Ramteke *et al*, 1991).

Previous researches have shown that the cell membrane protein component of such bacteria changed and this change may be limited to enhanced resistance for antimicrobial substances (Brown,1977; Harakeh *et al*, 1985). Bacterial resistance may also be acquired as a result of mutations in the bacterial chromosomes such mutations appear spontaneously as a result of damage to DNA the antibiotics provide the selection pressure for these drug-resistant mutants (Wray, 1986).

More often resistance phenomenon is plasmid (Extra chromosomal genetic material) mediated and transferred in nature resulting into its spread among the sensitive aquatic bacteria including coliforms (Pathak and Gopal,1994). Antibiotic resistant bacteria make their appearance as a result of exposure to aquatic environment, contaminated with faecal wastes from homeotherms by co-incidental co-selection of resistant factors (Grabow *et al*, 1975). ARI observed for the test strains significantly ranged from 0.014 to 0.023 exhibiting the broad-spectrum resistance for one to eight out of test antibiotics tested.

Table 7.5: Summary Bacteriological Quality of Drinking Water in Lucknow City

Sl.No.	*Areas*	*Locations*	*Total No. of Sample*	*Coliform* Contaminated Samples*	*Faecal Coliform* Contaminated Samples*	*Faecal Streptococci Contanminated Samples*
1.	RESIDENTIAL	Aliganj	5	2	1	1
2.		Vikas Nagar	5	1	0	0
3.		Indira Nagar	5	1	0	0
4.		Gomti Nagar	5	1	1	0
5.		Paper Mill Colony	5	1	1	0
6.		Nirala Nagar	5	2	2	2
7.		Triveni Nagar	5	0	0	1
8.		Rajajipuram	5	0	0	1
9.		Model House	5	1	1	0
10.		Ashiana	5	2	3	3
11.		Sadar Cantt.	5	0	1	2
	Total		**55**	**11/55 = 20%**	**10/55 = 19%**	**10/55 = 19%**
12.	COMMERCIAL	Charbagh	5	2	2	2
13.		Alambagh	5	2	2	0
14.		Chowk	5	1	2	1
15.		Hazratganj	5	1	1	0
16.		Hussainganj	5	0	0	1
17.		Aminabad	5	0	0	2
	Total		**30**	**6/30 = 20%**	**7/30 = 23%**	**6/30 = 20%**
18.	INDUSTRIAL	Amausi	5	1	1	1
19.		Talkatora	5	1	1	0
20.		Aishbagh	5	2	1	0
	Total		**15**	**4/15 = 27%**	**3/15 = 20%**	**1/15 = 7%**

* Samples with> 10 coliform/100 ml and/or >1 faecal coliform/100ml. As per Indian Standard for Drinking Water Specifications of Bureau of Indian Standard (BIS, 1996).

Although, the occurrence of indicator organisms of faecal pollution and incidence of multiple antibiotic resistance among bacterial population, particularly Gram-negative bacteria in drinking water of Lucknow city is surprising, it have well reported various other contaminated water sources (Ramtake *et al*, 1991). The microbial resistance to an antibiotic is a sign of potential health hazards.

Multidimensional worldwide development in most of areas has remitted into pollution in various components of Environment; water is also one of them. Municipal water and industrial waste have contaminated our drinking water. As we consumed the contaminated water, their contaminants indirectly affect our health. In this study, coliform, faecal coliform and faecal streptococci were isolated from drinking water of local areas in Lucknow city. The total bacterial load in drinking water was determined by multiple tube fermentation method and described as MPN/100ml which varies from <3/100ml. to >1100/100ml of coliforms and faecal coliforms and also faecal streptococci were observed to be <3/100ml to 460/100ml.

Table 7.6: Identification of Coliform Isolates by Biochemical Characterization

Test	*Bacterial Isolates*																
	1	*2*	*3*	*4*	*5*	*6*	*7*	*8*	*9*	*10*	*11*	*12*	*13*	*14*	*15*	*16*	*17*
Gram Reaction	–	–	–	–	–	–	–	–	–	–	–	–	–	–	–	–	–
Indole, 37°C	–	–	–	–	+	–	+	+	+	+	–	–	+	–	–	+	–
Indole, 44°C	–	+	–	–	+	+	+	–	+	+	+	+	+	+	+	–	–
MR	–	–	+	–	±	–	+	+	+	–	±	–	+	–	+	–	–
VP	+	+	+	–	+	+	+	–	–	+	–	+	–	+	±	+	–
Citrate	+	+	+	+	+	+	+	–	+	+	–	+	–	+	+	+	+
Urease	+	+	+	–	+	+	+	+	–	+	–	+	–	–	±	+	+
H_2S Production	+	+	–	–	–	+	+	–	–	–	±	–	–	–	–	–	±
Motility	+	+	–	–	+	+	+	–	+	–	+	+	+	–	+	–	–
Oxidase	+	+	–	–	–	–	+	+	–	–	–	–	–	–	–	–	–
Catalase	+	+	+	+	+	+	+	–	+	+	+	+	+	+	+	+	+
Gelatin lequifection	–	–	–	+	–	–	–	+	–	+	–	+	–	+	+	+	+

Bacteria Identified:

1. *Enterobacter cloacae*; 2. *Enterobacter cloacae*; 3. *Klebsiella pneumoniae*; 4. *Pseudomonas aeruginosa*; 5. *Enterobacter aerogens*; 6. *Enterobacter aerogens*; 7. *Enterobacter cloacae*; 8. *Proteus mirabilis*; 9. *Citrobacter koseri*; 10. *Klebsiella ozanae*; 11. *Escherichia coli*; 12. *Enterobacter aerogens*; 13. *Escherichia coli*; 14. *Klebsiella pneumoniae*; 15. *Eshcherichia coli*; 16. *Enterobacter aerogens*; 17. *Klebsiella pneumoniae.*

Table 7.7: Identification of Coliform Isolates by Biochemical Characterization

Test	*Bacterial Isolates*																
	18	19	20	21	22	23	24	25	26	27	28	29	30	31	32	33	34
Gram Reaction	–	–	–	–	–	–	–	–	–	–	–	–	–	–	–	–	–
Indole, 37°C	+	±	–	–	–	–	–	–	+	+	+	+	–	+	+	+	–
Indole, 44°C	+	+	–	–	–	–	–	–	+	–	+	–	–	+	+	+	+
MR	–	+	+	–	+	–	–	+	+	+	+	+	+	+	+	+	–
VP	–	–	+	–	±	±	+	±	–	–	–	–	–	–	–	–	+
Citrate	±	+	+	+	–	+	+	+	+	–	+	+	±	–	±	–	+
Urease	±	±	+	+	–	+	+	+	±	–	±	+	+	±	+	–	+
H_2S Production	–	–	+	–	–	–	–	±	–	–	–	–	+	–	–	–	–
Motility	+	+	+	+	+	–	–	+	+	+	–	+	–	+	–	–	–
Oxidase	–	+	–	+	–	–	–	–	–	–	–	–	–	–	–	–	–
Catalase	+	+	+	+	+	+	+	–	+	+	+	+	+	+	–	+	+
Gelatin lequifection	–	+	–	–	–	+	+	+	–	–	–	–	–	–	+	+	+

Bacteria Identified:

18. *Escherichia coli*; 19. *Citrobacter diversus*; 20. *Klebsiella pneumoniae*; 21. *Enterobacter cloacae*; 22. *Escherichia coli*; 23. *Klebsiella pneumoniae*; 24. *Klebsiella pneumoniae*; 25. *Citrobacter diversus*; 26. *Escherichia coli*; 27. *Escherichia coli*; 28. *Escherichia coli*; 29. *Citrobacter diversus*; 30. *Klebsiella pneumoniae*; 31. *Escherichia coli*; 32. *Citrobacter koseri*; 33. *Escherichia coli*; 34. *Klebsiella ozanae.*

Table 7.8: Antibiotic Resistance Among Coliform Bacteria from Drinking Water of Lucknow City

Strain No.	*Am 10μg*	*C 30 μg*	*Cf 10 μg*	*G 10 μg*	*K 30 μg*	*Nz 30 μg*	*Nx 10 μg*	*S 10 μg*	*T 30 μg*	*Tr 5 μg*	*ARI*
1	R			R	R	R	R	R	R	R	0.023
2	R			R	R	R	R	R	R	R	0.023
3	R	R		R	R	R			R	R	0.020
4	R				R	R			R	R	0.014
5	R				R	R			R	R	0.014
6	R		R		R	R	R	R	R	R	0.023
7	R			R	R	R		R	R	R	0.020
8	R				R	R		R	R	R	0.017
9	R				R	R			R	R	0.014
10	R	R			R	R			R	R	0.017
11	R	R			R	R		R	R	R	0.020
12	R	R			R	R		R	R	R	0.020
13	R			R	R	R	R	R	R	R	0.023
14	R	R			R	R			R	R	0.017
15	R	R			R	R			R	R	0.017
16	R	R			R	R			R	R	0.017
17	R				R	R	R	R	R	R	0.020
18	R	R			R	R		R	R	R	0.020
19	R	R			R	R		R	R	R	0.020
20	R	R			R	R		R	R	R	0.020
21	R	R			R	R		R	R	R	0.020
22	R		R	R	R	R	R	R	R	R	0.026
23	R	R			R	R		R	R	R	0.020
24	R	R			R	R		R	R	R	0.020
25	R	R			R	R		R	R	R	0.020
26	R	R			R	R		R	R	R	0.020
27	R				R	R	R	R	R	R	0.020
28	R					R	R	R	R	R	0.017
29	R	R				R	R	R	R	R	0.020
30	R					R	R	R	R	R	0.017
31	R				R	R		R	R	R	0.017
32	R	R			R	R		R	R	R	0.020
33			R		R	R			R	R	0.014
34	R	R			R	R		R	R	R	0.020

R: Resistant; ARI: Antibiotic resistance index.

Only Thirty-four (34) bacterial strains were isolated from hundred samples of water, collected from Lucknow. These bacterial strains were purified and identified by using different biochemical test. The coliforms and faecal coliforms were identified by IMViC test (Cowan and Steel, 1998) urease, catalase, oxidase test and motility were also performed for identification. Most of them are non-pathogenic or opportunistic pathogens. Isolated bacterial species are *Escherichia coli, Klebsiella sp., Citrobactor sp., Enterobacter sp. Pseudomonas sp.* and *Proteus sp.*

The isolated bacterial strains were showed a varying degree of antibiotic resistance with minimum inhibitory concentration. High percentage of isolates were observed to be resistant to Amoxycillin (Am), Kanamycin (K), Tetracycline (T) and Trimethoprim (Tr). Thirty-four isolates were observed to multiple antibiotic resistance (MAR). Maximum resistance (for 8 antibiotics) with 0.023 antibiotic resistance index (ARI) was shown by *Enterobater cloacal, Enterobacter aerogens* and *E. coli.*

The presence of multiple antibiotic resistant coliforms are harmful for health of human beings and the survey indicate that the drinking water of Lucknow city is highly affected by the faecal pollution.

Due to the faecal pollution of water human being faces various bacterial diseases, which causes the outbreaks and mass mortality. Some coliform bacteria are resided as commensally in gastrointestinal tract but sometimes they are also causing the diseases. The various strains of coliforms were isolated from drinking water. In these isolates, most of bacteria were found resistant to various antibiotics.

So, it is necessary that as far as possible, water sources must be protected from contamination by human and animal waste, which can contain a variety of pathogens, and parasites that will expose the community of pathogens and parasites that will expose the community to the risk of outbreaks and other infectious diseases.

References

APHA, 1995. *Standard Methods for Examination of Water and Wastewater.* American Public Health Association, 19th edition. Inc., New York, pp. 1170.

Armstrong, J.L., Shigeno, D.S., Calomiris, J.J. and Seidler, R.J., 1981. Antibiotic resistant bacteria in drinking water. *Appl. Environ. Microbiol.*, 42: 277–283.

BIS, 1996. *Indian Standard: Drinking Water Specification,* First Revision (3rd Reprint, October 1996). IS 10500–1991,Bureau of Indian Standard, New Delhi.

Brown, M.R.W., 1977. Nutrition depletion and antibiotic susceptibility. *Antimicrob. Chemother.*, 3: 198–201.

Cowan, S.T. and Steel, K.J., 1998. *Manual for Identification of Medical Bacteria.* The Cambridge University Press, London, 3rd Edn.

Grabow, W.O.K., Prozesky, O.W. and Burger, J.S., 1975. Behavior in a river and dam of a coliform bacteria with transferable and non-transferable drug resistance. *Water Res.*, 9: 777–782.

Harakeh, M.S., Berg, J.D., Hoff, J.C. and Matin, A., 1985. Susceptibility of chemostate grown *yersina enterocolitica* and *Klebsiella pneumonia* to chlorine dioxide. *App. Env. Microbiol.*, 45: 69–72.

Kelch, W.J. and Lee, J.S., 1978. Antibiotic resistance patterns of Gram-negative bacteria, isolated from environmental sources. *Appl. Env. Microbiol.*, 36: 450–456.

Mulamattathil, S.G. Esterhuysen, H.A. and Pretorius, P.J., 2000. Antibiotic resistant Gram–negative bacteria in a virtually closed water reticulation system. *Appl. Env. Microbiol.*, 88: 930–937.

Pathak, S.P. and Gopal, K., 1994. Antibiotic resistance and metal tolerance among coliform sp. from drinking water in a hilly area. *Env. Biol.*, 15(2): 139–147.

Pathak, S.P., Bhattcherjee, J.W. and Ray, P.K., 1993. Seasonal variation in survival and antibiotic resistance among various bacterial populations in a tropical river. *Gen. Appl. Microbiol.*, 39: 47–56.

Ramteke, P.W., Pathak, S.P. and Bhattacharjee, J.W., 1991. Isolation of antibiotic resistant coliform sp. from rural drinking water. *Environ. Bio.*, 12: 153–158.

Singh, D.K., 2004. Isolation, identification and characterization of coliform bacteria from drinking water in Lucknow city. *Dissertation Report*, pp. 1–75.

Trivedi, R.K. and Goel, P.K., 1984. *Chemical and Biological Methods for Water Pollution Studies.* Environmental Publication, Karad, pp. 1–251.

WHO, 1993. *Guidelines for Drinking Water Quality: Recommendations Microbiological Aspects* 2nd edn. Vol. I, WHO, Switzerland, Geneva.

WHO, 1999. *Guidelines for Drinking Water Quality*, 2nd edn. Vol. I, WHO, Switzerland, Geneva.

Wray, C., 1986. Some aspects of the occurrence of resistant bacteria in the normal animal flora. *Antimicrob. Chemother.*, 18 Suppl. C.: 141–147.

Chapter 8

Physico-chemical Parameters of the Benthic Environment of the Continental Slope of Bay of Bengal

Surajit Das, P.S. Lyla and S. Ajmal Khan

Introduction

Bay of Bengal, located in the northeastern part of the Indian Ocean, occupies an area of ~2.2x10^6 km^2 and is significantly influenced by both south-west (summer) and the north-east (winter) monsoons. It receives annual freshwater discharge of ~1.5x10^{12} m^3/yr (Subramanian,1993). This is accompanied by a sediment discharge of 1.4x10^9 tonnes/yr. Bay of Bengal showed a upwelling along a 40 km wide band along the southward due to the result of local long shore wind stress (Shetye *et al*, 1991), which disappears northward, overwhelmed by the enormous freshwater influx from Ganga- Brahmaputra rivers (Kumar and Ramesh, 2005). Cross-slope transfer of nutrient-rich deep oceanic water and local upwelling conditions along the continental margin and at the shelf edge may help to sustain the high productivity of shelf seas (Van Weering *et al* 1998). Studies on the physical, chemical and biological process at ocean margins- the shelf break and the continental slope- would help to determine the transport of material from the shelf to the deep sea (Flach *et al* 2002).

The diversity, distribution and relative abundance of different organisms in the marine environment are influenced by the physico-chemical properties of both the sediment and water. The hydrographical parameters of the biotopes are largely responsible for the biological productivity. The changing nature of water which, directly affects/alters the general texture and character of the sediments will in turn affect the biota. In view of the importance and involvement of the hydrographical parameters in various biological processes, investigations of these parameters have become mandatory for any environmental monitoring study. Knowledge on the role of sediment nutrients is thus useful in determining the sediment–water interaction which eventually affects the productivity of the overlying water.

In the present study, therefore, attempts were made to monitor hydrographical and sedimentological parameters of the benthic region of the continental slope of Bay of Bengal.

Materials and Methods

Samplings were carried out in the voyage of FORV *Sagar Sampada* (Ministry of Earth Sciences, Government of India). Sampling was carried out during cruise# 225 in April-May, 2004 from the area between latitude 10°36′N- 20°01′N and longitude 79°59′E- 87°30′E, covering 33 stations over 11 transects (three stations on each transect, each at the depth of ca. 200, 500 and 1000m) in the Bay of Bengal (Table 8.1 and Figure 8. 1).

Hydrography

Hydrographical parameters were obtained through sensors in all the stations using Sea-Bird 911 plus CTD instrument. For the analysis of dissolved oxygen the water samples were collected by CTD

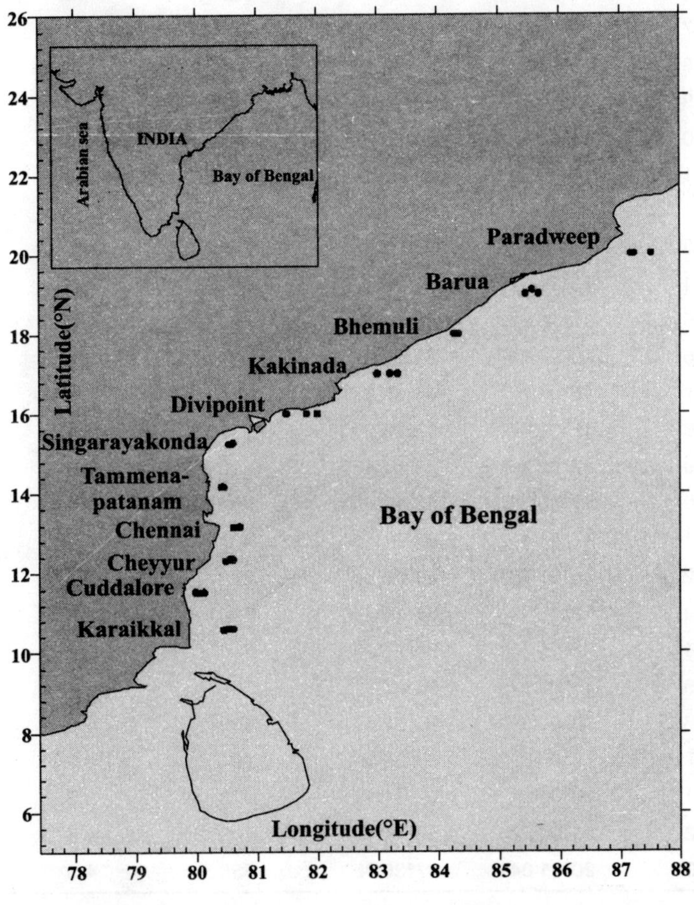

Figure 8.1: Study Area–Continental Slope of Bay of Bengal

rosette niskin bottles firing from the on board unit. Salinity and dissolved oxygen were expressed in psu and ml/l respectively.

Table 8.1: Location, Water Depth, Organic Carbon and Texture of Sediments for the Stations

Transects	*St. Nos.*	*Latitude (N)*	*Longitude (E)*	*Depth (m)*	*TOC (mg/g)*	*Sediment Texture*
Karaikkal	S48	10°34'96"	80°26'70"	248	2.5	Sandy clay loam
	S49	10°36'64"	80°30'67"	521	2.3	Sandy clay loam
	S50	10°36'16"	80°36'23"	950	4.5	Clay
Cuddalore	S51	11°31'82"	79°59'01"	214	2.6	Sandy clay
	S52	11°31'44"	80°02'16"	512	4.5	Clay
	S53	11°32'03"	80°07'55"	912	3.9	Silty clay
Cheyyur	S54	12°18'81"	80°29'35"	216	3.0	Sandy loam
	S55	12°21'68"	80°33'49"	490	3 5	Silty clay
	S56	12°21'34"	80°36'10"	937	4.0	Silty clay
Chennai	S57	13°09'78"	80°36'66"	225	4.3	Silty clay
	S58	13°09'95"	80°41'99"	516	3.5	Silty clay
	S59	13°10'21"	80°43'18"	970	3.8	Sandy clay
Tammenapatanam	S60	14°10'65"	80°24'97"	228	4.5	Clay
	S61	14°10'99"	80°26'37"	477	5.5	Clay
	S62	14°10'23"	80°27'04"	1000	4.9	Silty clay
Singarayakonda	S63	15°15'19"	80°31'68"	240	4.6	Clay
	S64	15°15'32"	80°33'47"	507	4.8	Clay
	S65	15°16'33"	80°35'82"	960	5.5	Clay
Divipoint	S66	16°00'09"	81°29'57"	223	4.9	Clay
	S67	16°04'70"	81°49'67"	502	5.5	Clay
	S68	16°00'67"	82°03'29"	917	4.9	Silty clay
Kakinada	S69	17°00'16"	83°01'86"	220	5.5	Clay
	S70	17°00'07"	83°12'70"	500	4.8	Clay
	S71	17°01'45"	83°21'18"	901	5.3	Clay
Bhemuli	S72	18°01'49"	84°15'62"	222	4.7	Clay
	S73	18°01'93"	84°16'26"	550	5.3	Clay
	S74	18°01'13"	84°20'28"	850	5.0	Clay
Barua	S75	19°03'37"	85°25'81"	260	4.8	Clay
	S76	19°06'37"	85°32'55"	500	4.8	Clay
	S77	19°05'63"	85°39'03"	994	4.0	Clay
Paradweep	S81	20°05'29"	87°11'14"	273	5.2	Clay
	S82	20°04'32"	87°12'19"	503	4.8	Clay
	S83	20°01'04"	87°30'95"	958	4.3	Silty clay

Sedimentology

Sediment samples were collected by Smith McIntyre Grab with $0.2m^2$ coverage area. Temperature of the sediment samples were measured by a centigrade thermometer as soon as the grab was hauled up and expressed in °C. pH of the sediments was measured by a soil pH meter (Eutech Instruments, Singapore). Sediment water content was estimated by weight loss of wet sediment samples dried at 60°C and was expressed in percentage. Porosity was determined with the following equation: (wc/1.02)/{[(1-wc)/2.64] + wc/1.02}, where wc is [(wet sediment weight–dry sediment weight)/wet sediment weight] (Danovaro *et al* 1999). The texture of sediments was analyzed by the combined method of sieving and pipette analysis according to the procedure given by Krumbein and Pettijohn (1938). Sand, silt and clay fractions were expressed in percentage. The values are then plotted on a trigon plot and the textural characteristics of the samples were determined (Folk, 1968). Total Organic Carbon content (TOC) was estimated by chromic acid oxidation method followed by titration with ammonium ferrous sulfate (El Wakeel and Riley,1956) and expressed in mg/g dry sediment.

Statistical Analysis

The correlation coefficients of various physico-chemical parameters in relation to each other were calculated. Multidimensional scaling (MDS) and Principal Component Analysis (PCA) were calculated with the help of computer software (Primer V5) to group the stations based on the similarity.

Results and Discussion

The Indian continental margin is an important element in the evolutionary history of the break-up of the Indian Plate from Gondwanaland. The break-up between the east coast of India and Antarctica led to the development of the eastern continental margin of India followed by its subsidence and sediment deposition in the basins. The Bay of Bengal region is covered with pre- and post- collision sediments. The nature of crust beneath the Bay of Bengal is still controversial and it has been argued that either it is continental or oceanic type (Sridhar *et al* 2006). The Arabian Sea and the Bay of Bengal form the arms of the northern Indian Ocean. But, whereas the former has been studied in considerable detail, the later still remains as one of the least studied areas of the Indian Ocean.

The physico-chemical parameters recorded at different depths of continental slope in Bay of Bengal are presented in Table 8.1 and Figure 8.2.

The correlation co-efficient matrix of the physico-chemical parameters is presented in Table 8.2.

The sediment overlying water salinity ranged between 34.73 (st. 50, off Karaikkal, 1000m) and 35.02 psu (st. 52, off Cuddalore, 500m). Average salinity was found to be higher in 500m stations (35.01±0.005 psu) followed by 200m (34.94±0.04 psu) and 1000m stations (34.92±0.07 psu). Dissolved oxygen of the bottom water varied between 0.12 (st. 72, off Bhemuli, 200m) and 2.46 ml/l (st. 50, off Karaikkal, 1000m). Higher DO was observed in the 1000m depth stations (1.06±0.61 ml/l) followed by 500m (0.29±0.14 ml/l) and 200m stations (0.27±0.13 ml/l). DO showed highly significant positive correlations with depth, salinity, sediment temperature and silt fraction of sediment.

The annual freshwater discharge in Bay of Bengal is ~$1.5x10^{12}$ m^3/yr (Subramanian,1993). This huge influx of freshwater leads to a significant decrease in salinity and increase in sea surface temperature during summer causing a strong stratification. Mixing is further prevented due to the presence of low salinity water in the upper layer. The maximum salinity was found at a depth of about 500m, as highly saline waters moves into the Bay of Bengal from the Indian Ocean (Venkataraman,2003). In the present study, therefore, salinity was negatively correlated with dissolved oxygen which was positively correlated with depth.

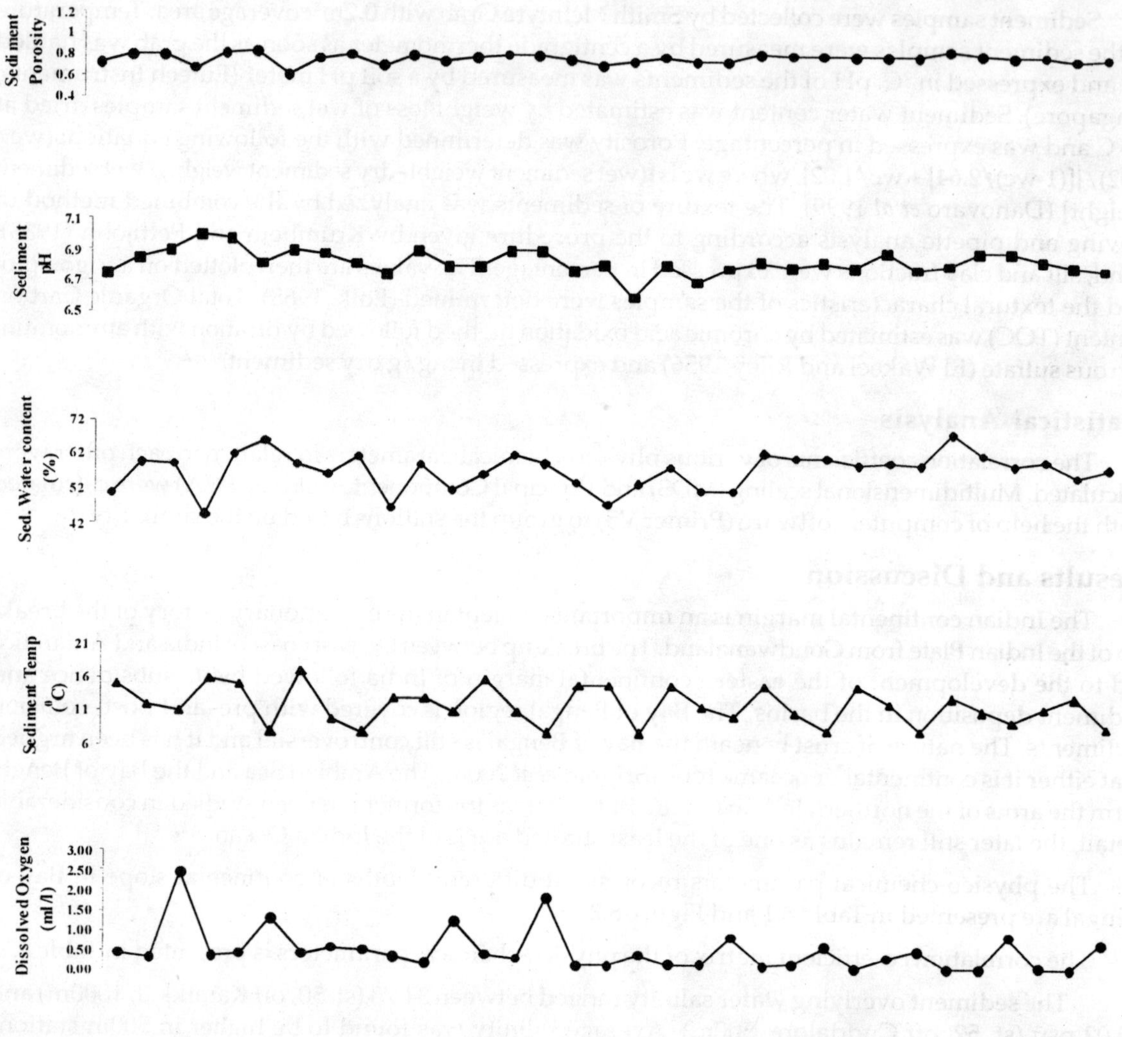

Figure 8.2: Physico-chemical Parameters of the Continental Slope of Bay of Bengal

Table 8.2: Correlation Co-efficient Matrix (n = 33) for Physico-chemical Parameters Recorded During the Study

	Depth	*Salinity*	*Dis. Oxy.*	*Sed. Temp.*	*Sed. pH*	*Water Cont.*	*Porosity*	*Sand %*	*Silt %*	*Clay %*	*TOC*
Depth	1.00										
Salinity	-0.18	1.00									
Dis. Oxy	0.68***	-0.65***	1.00								
Sed. Temp	-0.85***	0.03	-0.48**	1.00							
Sed. pH	-0.09	-0.04	0.18	0.21	1.00						
Water cont.	0.14	0.02	0.06	-0.27	0.25	1.00					
Porosity	0.26	-0.01	0.08	-0.43*	0.09	0.78***	1.00				
Sand per cent	-0.34*	0.20	-0.11	0.43*	0.21	-0.29	-0.58***	1.00			
Silt per cent	0.51**	-0.29	0.46**	-0.45**	-0.04	-0.02	0.06	-0.48**	1.00		
Clay per cent	-0.04	0.02	-0.25	-0.11	-0.20	0.33	0.58***	-0.70***	-0.29	1.00	
TOC (mg/g)	0.14	-0.08	-0.08	-0.24	-0.30	0.26	0.40*	-0.73***	0.11	0.71***	1.00

Level of significance: ***$P < 0.001$; **$P < 0.01$; *$P < 0.05$.

The bottom water in Bay of Bengal is exposed to the waters derived from the West Australia which is nutrient-rich but oxygen-poor waters. Edmond *et al.* (1979) and Broecker *et al.* (1980) described a nutrient-rich benthic layer within the Bay of Bengal and noted a sharp decrease in oxygen concentration at the sediment-water interface and a deep oxygen maximum layer is produced by spreading of slightly better ventilated West Australian Basin water over the denser nutrient-rich benthic layer. There are two paths for bottom water to ventilate the floor of the Bay of Bengal, both derived from waters of the West Australian Basin: an eastern path directly from the West Australian Basin and the overflow path via the Central Indian Basin (Gordon *et al* 2002). West Australian Basin deep water influence the sea floor shallower than about 3400m, and the Central Indian Basin deep water ventilates the sea floor in the southern area.

Cowie *et al.* (1999) indicated that the oxygen availability is an important factor in determining sedimentary organic matter content. Oxygen is energetically the most preferred electron acceptor for microbial degradation and is, therefore, consumed first in the diagenetic sequence (Canfield 1993 and Martin and Sayles 2004). If there is a sufficient supply of degradable organic matter, oxygen is depleted within the uppermost millimeters to centimeters of the sediment. But, in the present study, there was no apparent relationship found between the total organic carbon contents of the sediments and bottom-water oxygen concentrations coinciding with the result of Calvert *et al.* (1992).

Sediment temperature varied between 8°C (st. 53, off Cuddalore, 1000m) and 17°C (st. 54, off Cheyyur, 200m). The sediment temperature showed a regular trend of decreasing from ca. 200m to ca. 1000m. Temperature showed negative correlation with depth, porosity and silt, and positive correlation with sand fraction of sediment.

Sediment pH varied between 6.6 (st. 65, off Singarayakonda, 1000m) and 7.0 (st. 51, off Cuddalore, 200m). The variations in pH from different stations were found to be very less.

The sediment water content varied between 44.73 per cent (st. 51, off Cuddalore, 200m) and 68.54 per cent (st. 75, off Barua, 200m). The water content in the 200m stations showed higher values

(56.66±7.07 per cent) followed by 1000m (58.54±4.32 per cent) and 500m (58.16±4.76 per cent). It showed a highly significant positive correlation with the porosity of sediment.

Porosity varied between 0.61 (st. 54, off Cheyyur, 200m) and 0.84 (st. 75, off Barua, 200m). Higher porosity was found in 500m and 1000m depth stations (0.78±0.03) followed by 200m stations (0.75±0.06). Higher the porosity higher the sediment water content which is positively correlated.

The arrangement of various particles in sediments creates a complex porous system filled with continuously changing ratios of water and air (Hedges and Oades,1997). Surface marine sediments are less dense and the porosities of fine-grained deposits range from 0.7-0.9 in the upper few centimeters and approach values near 0.5-0.6 at depths of a meter or more (Berner,1980). In the present study also the porosity was found in that range. The higher porosities and looser packing of fine-grained minerals in surface marine sediments appear to derive largely from continuous buoying of the component solids by pore waters favoring dispersion (Hedges and Oades,1997).

The porosity affects the diffusion and transfer of dissolved materials between the sediment and the overlying water. Diffusion of dissolved substances within the sediment takes place only in the pores and thus is approximately linearly depended on porosity (Avnimelech *et al* 2001).

The texture of the sediments were found mostly clayey, other types of sediments recorded were silty clay, sandy clay, sandy clay loam (Table 8.1). The sand fraction in the sediments from the different stations varied between 3 per cent and 62 per cent. The silt percentage ranged between 10 per cent and 53 per cent and the clay fractions varied between 12 per cent and 78 per cent. Sand fraction negatively correlated with depth, and it was found that the deeper stations showed less sand content.

The texture of the sediment in the present study was found to be clay in the northern stations and silty clay in the southern stations. The result coincides with the earlier report of the Indian deep sea venture. Ansari (2000) and Valsangkar and Ambre (2000) reported the texture of the sediment to be fine clay and silty clay from the Central Indian Basin of 5000 to 5200 m depth. They emphasized that the greater depths of the Indian Ocean Basin sediments were the result of the travel of the terrigenous materials in suspension quite far from the coast before being deposited, which may also be attributed in the case of Bay of Bengal, because it is a unique basin receiving considerable amount of freshwater and sediment discharge from major rivers *viz.*, Ganges, Brahmaputra, Godavari, Krishna, Mahanadi and Cauvery. Among these rivers, the Ganga-Brahmaputra riverine system transports more than 1000 million tons of suspended sediments to the Bay, highest among all the global rivers (Madhupratap *et al* 2003 and Chauhan 2003). In support, Vaz *et al.* (1996) reported the sediment characteristic from continental slope at 495m water depth off Madras, Bay of Bengal. The upper (0-94cm) sediment was found to be sandy clay type, while the lower part was clayey sand. Since, the rate of sedimentation within the Bay of Bengal is expected to vary due to variable input from fluvial (organic and undecompressed matter) sources (Chauhan, 2003), the sediment texture determined in the present study varied from southern to northern transects. Northern transects, in addition to the Ganga and Brahamaputra riverine system, receive the sediments of the rivers Mahanadi, Godavari and Krishna, the major rivers debouching and draining from the peninsular India. In contrast, the southern Bay of Bengal transects receive less drainage and discharges from the river Cauvery and a part of the river Krishna.

Organic matter in continental margin sediments has an impact on environmentally important chemical processes far out of proportion to its relative abundance. It is an active scavenger of metals and other types of pollutants and the burial of organic carbon in sediments (Ransom *et al* 1998). Thus, the cycling of organic compounds between the hydrosphere and lithosphere influence levels of atmospheric CO_2 (Broecker *et al* 1979).

Total organic carbon in the present study varied in relation to the texture of the sediment and varied from 2.3 mg/g (st. 49, off Karaikkal, 500m) to 5.5 mg/g dry sediment weight basis (st. 69, off Kakinada, 200m). In the 1000m station TOC was found to be higher (4.5±0.5 mg/g) followed by 500m (4.4±0.9 mg/g) and 200m stations 4.2±1.0 mg/g).

Continental margins and deltaic settings are important sites for organic carbon accumulation because they receive large amounts of organic matter from both terrestrial and marine sources (Berner, 1982).

Rivers are the main source of terrigenous sediment delivered to continental margins and thus, exert a major control on coastal evolution and sequence development. Development of the Ganges-Brahmaputra River delta began ca. 11000 yr B.P., when rising sea level flooded the Bengal basin, thereby trapping most of the river's discharge on the inner margin. Chronostratigraphic study calculated the rates of sediment storage revealed that $5x10^{12}$ m^3 of sediment was stored in the Bengal basin from ca. 11000 to 7000 yr B.P., which corresponds to a mean load of $2.3x10^9$ t/yr. In comparison, modern sediment load of the Ganges-Brahmaputra is >1000 million tones/yr, ranking it first among the world's rivers and underscoring the significance of a two-fold increase sustained over 4 k.y (Goodbred and Kuehi, 2000).

It is well known fact that in sediments the organic matter is mainly attached to the fine grained fraction. In general a strong positive correlation was found between the C_{org} concentration and the fraction of clay in the present study as was evident by (Rao and Veerayya, 2000 and De Hass *et al* 2002). Seralathan and Seetaramaswamy (1979) attributed the increase of organic matter in fine particles is the similarity in the settling velocity of both organic constituents and fine particles. Sand percentage as usual showed negative correlation with organic carbon coinciding with the result of Boon and Duineveld (1996). But no correlation was found with depth as supported by the result of the slope sediment of NE Atlantic (Flach and Heip, 2002).

The lithogenic fluxes are greater in the Bay and increase in fluxes are associated with increase in river discharge (Ittekkot *et al*, 1991 and Ramaswamy and Nair, 1994). It may be that biogenic matter is more rapidly scavenged along with those of terrigeneous origin and ballast the material resulting in faster sedimentation to the deep (Ittekkot *et al*, 1992). Thus, it may be fluvial matter [organic and undecomposed matter (Sen Gupta *et al.*, 1977)] that might be contributing more to the fluxes.

In the Bay of Bengal organic carbon is ballasted into the deep by the high lithogenic flux from rivers, which form aggregates with the former (Ittekot *et al* 1991). This may be attributed in the low vales of organic carbon in the continental slope of Bay of Bengal. Beside, the lower euphotic layer productivity in this bay also contributed less carbon. Further, the rate of sedimentation in the Bay of Bengal varied due to variable input from fluvial sources (Chauhan, 2003), which would lead to a much greater organic degradation at the sediment-water interface. Hence, the low organic carbon contents in the present study may be due to higher residence time of particles in the water column at greater depths and longer oxygen exposure time at the sediment-water interface in the Bay of Bengal.

All data derived from the sediment analyses, together with depth, were ordinated by a correlation based principal component analysis (PCA) (Figure 8.3). In non-metric multidimensional scaling (MDS), the Bray-Curtis similarity was used to construct a map in which those having more similarity were placed near and samples having lower similarity far away (Figure 8.4).

An ordination by a PCA of the sediment variables and depth showed a clear distinction. The first component of the total variance of the data, was the main one responsible for the separation of the sites. Higher positive values along the first component were related to sediment temperature, water content, sand fraction and water salinity, whilst higher negative scores along the second component were due to clay and silt percentage, sediment water content and water salinity (Figure 8.3).

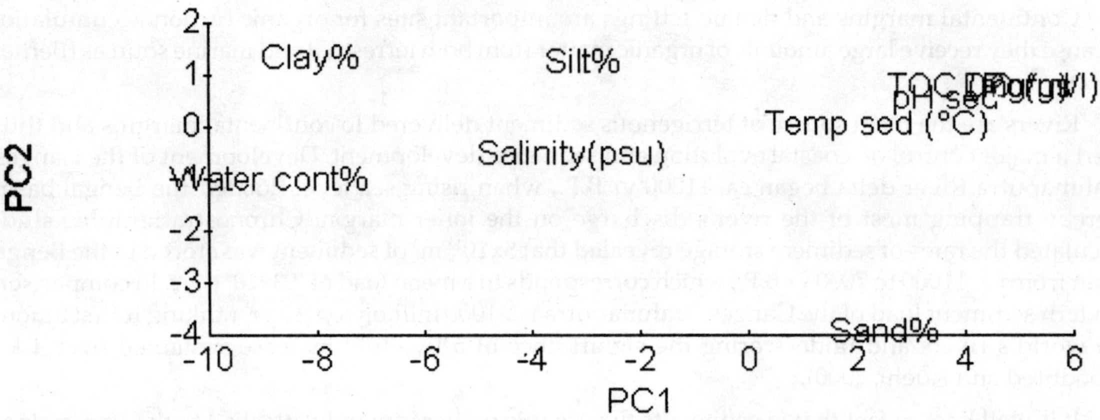

Figure 8.3: Principal Component Analysis (PCA) of the Physico-chemical Parameters

The non-metric multidimensional scaling (MDS) grouped the stations based on the similarity. There were two clusters formed in the figure. First cluster consists of 29 stations and second cluster consists of rest 4 stations. Second cluster differentiated from the first cluster by the location. St. 48, 49, 51 and 54 of second cluster were all from the southern part of Bay of Bengal, where the stations (off Karaikkal, Cuddalore, Cheyyur) were found to be differentiated by their physico-chemical parameters.

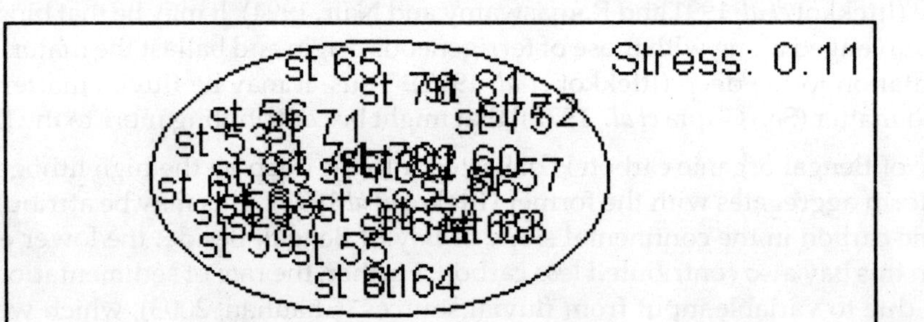

Figrue 8.4: Non-metric Multidimensional Scaling (MDS) of the Stations

Conclusion

A number of models or budgets have been constructed previously to characterize the biological cycling of organic matter in Bay of Bengal. Sediment community respiration is inversely related to depth, and therefore correlates directly with temperature, biomass and detrital organic matter food supplies. The continental slope sediments may be globally significant sites of benthic carbon remineralization despite their relatively small area, as rates of organic matter oxidation are greater than those found in the deep sea. In view of this, the present study on the physico-chemical parameters will help to ascertain the temporal changes of these parameters which will further go a long way to assess the benthic productivity of the continental slope of Bay of Bengal.

Acknowledgements

The authors are thankful to the Director, Centre of Advanced Study in Marine Biology for the encouragement, the authorities of Annamalai University for the facilities and the Centre for Marine Living Resources and Ecology, Ministry of Earth Sciences, Government of India, Kochi for financial support.

References

Ansari, Z.A., 2000. Distribution of deepsea benthos in the proposed mining area of Central Indian Basin. *Mar. Georesource. Geotechnol.*, 18: 201-207.

Avnimelech, Y., G. Ritvo, L. E. Meijer, M. Kochba, 2001. Water content, organic carbon and dry bulk density in flooded sediments. *Aquacult. Engineer.*, 25: 25-33.

Berner, R.A., 1980. Early diagenesis: A theoretical Approach. Princeton University Press. Princeton, NJ.

Berner, R.A., 1982. Burial of organic carbon and pyrite sulfur in the modern ocean: Its geochemical and environmental significance. *Am. J. Sci.*, 282: 451-473.

Boon, A.R. and G.C.A. Duineveld, 1996 Phytopigments and fatty acids as molecular markers for the quality of near-bottom particulate organic matter in the North Sea. *J. Sea Res.*, 35: 279-291.

Broecker, W., R. Toggweiler and T. Takahashi, 1980. The Bay of Bengal- a major nutrient source for the deep Indian Ocean. *Eart. Planet. Sci. Lett.*, 49: 506-512.

Broecker W.S., T. Takahashi, H.J. Simpson and T.H. Peng, 1979. Fate of fossil fuel, carbon dioxide, and the global carbon budget. *Science*, 206: 409-418.

Calvert, S.E., R.M. Bustin and T.F. Pedersen, 1992. Lack of evidence for enhanced preservation of sedimentary organic matter in the oxygen minimum of the Gulf of California. *Geol.*, 20: 757-760.

Canfield, D.E., 1993. Organic matter oxidation in marine sediments. In: 'Interactions of C, N, P, and S in Biogeochemical Cycles' (R. Wollast, L. Chou and F. Mackenzie Eds.). Pp. 333-363. NATO AS1 Ser. Springer, Berlin.

Chauhan, O.S., 2003. Geochemistry of ferromanganese micronodules and associated Mn and trace metals diagenesis at high terrigenous depositional site of middle fan region, Bay of Bengal. *Deep Sea Res.* II, 50: 961-978.

Cowie, G.L., S.E. Calvert, T.F. Pedersen, H. Schulz and U. von Rad, 1999. Organic content and preservational controls in surficial shelf and slope sediments from the Arabian Sea (Pakistan margin). *Mar. Geol.*, 161: 23-38.

Danovaro, R., D. Marrale, N. Della Croce, P. Parodi and M. Fabiano, 1999. Biochemical composition of sedimentary organic matter and bacterial distribution of in the Aegean Sea: trophic state and pelagic-benthic coupling. *J. Sea. Res.*, 142: 1-13.

De Haas, H., T.C. E. van Weering and H. de Stigter, 2002. Organic carbon in shelf seas: sinks or sources, processes and products. *Cont. Shelf Res.*, 22: 691-717.

Edmond, J., S. Jacobs, A. Gordon, A. Mantyla, and R. Weiss, 1979. Water column anomalies in dissolved silica over opaline pelagic sediments and the origin of the deep silica maximum. *J. Geophys. Res.*, 84: 7809-7826.

El Wakeel, S.K. and J. P. Riley, 1956. The determination of organic carbon in marine muds. *J. du. conseil permanent Intl. Poul. Exploration de la mer.*, 22:180- 183.

Flach, E., A. Muthumbi and C. Heip, 2002. Meiofauna and macrofauna community structure in relation to sediment composition at the Iberian margin compared to the Goban Spur (NE Atlantic). *Prog. Oceanogr.*, 52: 433-457.

Folk, R.L. 1968. Petrology of sedimentary rocks. Hemphills, Austin, Texas.

Goodbred, S.L. and S.A. Kuehl, 2000. Enormous Ganges-Brahmaputra sediment discharge during strengthened early Holocene monsoon. *Geol.*, 28: 1083-1086.

Gordon, A.L., C.F. Giulivi, T. Takahashi, S. Sutherland, J. Morrison and D. Olson, 2002. Bay of Bengal nutrient-rich benthic layer. *Deep Sea Res.* II, 49: 1411-1421.

Hedges, J.I. and J.M. Oades, 1997. Comparative organic geochemistries of soils and marine sediments. *Org. Geochem.*, 27(7/8): 319-361.

Ittekkot, V., R.R. Nair, S. Honjo, V. Ramaswamy, M. Bartsch, S. Manganini and B.N. Deșai, 1991. Enhanced particle fluxes in Bay of Bengal induced by injection of freshwater. *Nature*, 351: 385-387.

Ittekkot, V.B., B. Haake, M. Bartsch, R.R. Nair and V. Ramaswamy, 1992. Organic carbon removal in the sea: the continental connection. Upwelling Systems: Evolution since the early Miocene. Geological Society of London, Special Publication No. 64, pp. 167-176.

Krumbein, W.C. and F.J. Pettijohn, 1938. Manual of sedimentary petrography. Applenton Century-Crofts, New York.

Kumar, S. and R. Ramesh, 2005. Productivity in the Bay of Bengal using the ^{15}N tracer: Implications to the global carbon cycle. *Indian J. Mar. Sci.*, 34: 153-162.

Madhupratap, M., M. Gauns, N. Ramaiah, S. Prasanna Kumar, P.M. Muraleedharan, S.N. de Sousa, S. Sardessai and U. Muraleedharan, 2003. Biogeochemistry of the Bay of Bengal: physical, chemical and primary productivity characteristics of the central and western Bay of Bengal during summer monsoon 2001. *Deep Sea Res.* II, 50: 881-896.

Martin, W.R. and F.L. Sayles, 2004. Organic matter cycling in sediments of the continental margin in the northwest Atlantic Ocean. *Deep Sea Res.* I, 51: 457-489.

Ramaswamy, V. and R.R. Nair, 1994. Fluxes of material in the Arabian Sea and Bay of Bengal- sediment trap studies. *Proceed. Ind. Acad. Sci.*, 103: 189-210.

Ransom, B., D. Kim, M. Kastner and S. Wainwright, 1998. Organic matter preservation on continental slopes: importance of mineralogy and surface area. *Geochim. Cosmochim. Acta*, Abstract, p. 39.

Rao, B.R. and M. Veerayya, 2000. Influence of marginal highs on the accumulation of organic carbon along the continental slope of western India. *Deep Sea Res.* II, 47: 303-327.

Sen Gupta, R., S.N. De Sousa and T. Joseph, 1977. On nitrogen and phosphorous in the western Bay of Bengal. *Indian J. Mar. Sci.*, 6: 107-110.

Seralathan, P. and A. Seetaramaswamy, 1979. Geochemistry of modern deltaic sediments of the Cauvery river, East coast of India. *Indian J. Mar. Sci.*, 16: 31-38.

Shetye, S.R., S.S.C. Shenoi, A.D. Gouviea, G.S. Michael, D. Sundar and G. Nampoothiri, 1991. Wind driven coastal upwelling along the western boundary of the Bay of Bengal during southwest monsoon. *Cont. Shelf Res.*, 11: 1397-1408.

Sridhar, A.R., A. Pradeep Kumar, Uma Shankar and N.K. Thakur, 2006. Presence of illites in Bay of Bengal- An analysis of the sample obtained from GEODROME. *Curr. Sci.*, 90(4): 574-578.

Subramanian, V. 1993. Sediment load of Indian rivers. *Curr. Sci.*, 64: 928-930.

Valsangkar, A.B. and N.V. Ambre, 2000. Benthic disturbance and impact experiment in the Central Indian Ocean Basin. *Mar. Georesour. Geotechnol.*, 18: 189-200.

Van Weering, T.C.E., I.N. McCave and I.R. Hall, 1998. Ocean margin exchange (OMEX I) benthic processes study. *Prog. Oceanogr.*, 42: 1-4.

Vaz, G.G., P. Vijaykumar and B.L. Rao, 1996. Glaucony in a core from upper continental slope off Madras, Bay of Bengal. *Indian J. Mar. Sci.*, 25: 204-209.

Venkataraman, K., 2003. Marine ecosystems of India. In: 'Natural aquatic ecosystems of India, Thematic Biodiversity Strategy and Action Plan' (K. Venkataraman Ed.).Pp 57-114. The National Biodiversity Strategy Action Plan, India.

Chapter 9

Role of Limno-chemical Factors on Phytoplankton Productivity

N. Shiddamallayya and M. Pratima

Introduction

Primary production of organic matter provides two of the most essential components for all biological and chemical processes in aquatic ecosystem. It supplies energy for all the endothermic reactions and power for all the reducing reactions, which constitute most of the fundamental process in an ecosystem. Planktonic organisms form the first trophic level of the water body. It is to be a good index for water quality. There is a great variation in the composition of plankton in different regions and at different depths and also at different seasons at different hours probably even in different years. The conditions that lead to maxima and minima, as well as to minor fluctuations in abundance of any particular plant or animal are complex in their physical, chemical and biological aspects. An indicator species is the one chosen to represent conditions in the water body because of sensitive to changes in the environment.

Based on the distribution pattern of the plankton, the quality of the aquatic environment can be assessed. It is commonly observed that, the surface distribution of phytoplankton and zooplankton are often patchy in character. It is observed that great swarms of zooplankton may consume the phytoplankton to such an extent that, they create discontinuous distribution of plankton. The patchy distribution of phytoplankton may also be due to the differential rates of multiplication, which might induce similar effects. The fast growing phytoplankton produces dense blooms, where they are more abundant in water at a favorable time. Slow growing zooplankton might graze upon this and by the time population becomes heavy, the blooms of phytoplankton would have crossed its peak because of exhaustion of surface nutrients. It seems probable that, these mechanisms may play a role in producing the discontinuities at various times and places.

The phytoplankton is an important component of all aquatic ecosystem and it consists of primary producers with photosynthetic activity, that constitute the first step in the carbon cycle of the water body. Phytoplankton is a community of algae in which the population of individual species may vary in density.Often one or small groups of the species may be dominant, dependent upon the nutritional status of the water body.The wide distribution of phytoplankton and their frequent abundance accounts for the great importance of the phytoplankton as a basic food source in the food web of the aquatic ecosystems. The phytoplankton production is dependent on the availability of solar energy, micro and macronutrients as well as promotive organic substances.

In India, a considerable amount of work has been done on different types of fresh water bodies such as lakes, ponds, pools, reservoirs, and temple tanks by Ganapati (1943), Chacko and Krishnamoorthy (1954), George (1962) and Sreenivasan (1964 and 1969) in Tamil Nadu. Khan and Siddiqui (1971) have opined that, seasonal fluctuations in gross and net primary production values were quite apparent in surface water of fishpond, Aligarh. Gopal *et al.* (1978) have studied the primary production of different types of ponds at Bhagalpur. Seenayya and Zafar (1979) have reported the phytoplankton production in Mir Alam lake of Hyderabad. Palaniappan *et al.* (1981) have studied the two rock pools of Salum and revealed that the primary productivity of the rock pools was dependent on the amount, duration and frequency of rainfall and also on the location of the water body.

Ayyappan and Gupta (1985) have studied the primary production in Ramasamudra at Mangalore and described that, the primary production was controlled by several hydrochemical factors. Valecha and Bhatnagar (1989) have studied the production of phytoplankton in a eutrophic lake of Bhopal. Ravindra Singh (1990) has studied the north tank and south tank of Jamalpur, Munger to analyse the significant correlation between certain physico-chemical parameters and primary production of phytoplankton. Bais *et al.* (1997) have investigated nitrite and phosphate as the main factors for the control of phytoplankton production respectively in Sagar lake and Military engineering lake, Sagar. Singh (1998) has studied the primary production in relation to physico-chemical nature of the water body and annual fish catch in Gobindsagar reservoir of Himachal Pradesh. Naganandini and Hosmani (1998) have reported the ecology of certain inland water bodies in Mysore district. Singh and Singh (1999) have conducted a comparative study on the phytoplankton primary production of river Ganga and pond at Patna, Bihar. Prakasam and Joseph (2000) have analysed the water quality of Sasthamcotta lake Kerala in relation to primary productivity and pollution from anthropogenic sources. Jeyachandran and Krishnan (2001) have conducted a short-term study on certain physico-chemical factors and primary productivity at Viraganur dam, Madhurai. Das (2002) has studied the phytoplankton primary production in nine selected reservoir of Andhra Pradesh. Sukumaran (2003) has estimated the primary production of Lalbagh tank in Bangalore and noted the production was higher in the surface than the bottom. Rajkumar (2004) has made a review on the quantum of phytoplanktonic primary production of a polluted fresh water pond of Krishnan Anaikattikulum in Pollachi, Coimbatore. Mali and Gajaria (2004) have made an assessment of primary productivity and hydrobiological characterisation of a fish culture pond in Gujarat.

Study of primary production in water bodies of Northern part of Karnataka was limited to Gulbarga region only. Gaddad *et al.* (1983), Vijaykumar and Paul (1992 and 1994), Janakiram and Jayaraj (1996), Angadi *et al.* (1999) Vijay Kumar *et al.* (2000) and Eshwarlal and Angadi (2002) have conducted hydrobiological studies of some water bodies of Gulbarga district. However, no reports of role of limno-chemical factors on phytoplankton productivity in the water bodies of Bidar are available. Hence, two water bodies of Bidar district have been selected for the study.

Bidar, the head quarter of the district lies at 17° 55′ North latitudes and 77° 32′ East longitudes, located 551 meters above the mean sea level. The Papnash pond is situated at the Western side of the city. It covers an area of 0.25 square kilometer. East and West boundaries of pond are flanked by hills; fields demarcate Northern and Southern boundaries. *Acacia, Dalbergia* and other vegetation cover the Eastern and Western sides of the pond. The pond is under the cover of large population of *Ipomea fistulosa* Mart. The mean depth of pond is 2.1 meter with a maximum depth of 4 meter during the rainy season and minimum being 0.75 meter during dry season. This pond is exploited by the surrounding people for the various domestic activities, such as washing of clothes and bathing of cattle. Further this water body is also used for agriculture.

Bhalki is one of the towns in Bidar district and lies at 18° 03′ North latitude and 77° 12′ East longitudes, located 590 metres above the mean sea level. Bhalki tank is situated at the Northwestern side of the city. The tank covers an area of 3.18 square kilometer. The approach road from Bhalki to Siddapur Wadi is marking the shoreline of Southwest. Other sides like Northeastern side of the tank, is demarked with field, Northwest by dry land and Southeast by the city. More than half of the water body of the tank and the three sides of the shorelines has been covered with *Ipomoea fistula* Mart. The water level ranged from 1.2 meter in summer to 4.35 metres during rainy season. The mean depth of the tank during the study period was 2.8 metres. Southwest shoreline has two out lets, which are used to supply the water for the agricultural fields. Further, the Bhalki tank water is also used for fisheries, and for other domestic activities like bathing of cattle, washing of clothes, disposing of anthropogenic wastes and cleaning of vehicles.

Materials and Methods

Collection of Water Samples

The surface water samples were collected once in a month from both the water bodies of two sites, A and B of Papnash pond, Bidar and A and B of Bhalki tank, using wide mouthed clean Iodine treated Poly Vinyl Chloride container for the study of various limno-chemical factors. All collections and observations were made between 8.30 a.m. to 10.00 a.m. through out the study period. The collected water samples brought to the laboratory for the estimation of various parameters. Standard methods were employed for the estimation of various parameters such as pH, dissolved oxygen, free carbon dioxide, alkalinity, hardness, calcium, magnesium, chloride, nitrite, phosphate, silicon, sulphate, total solids and organic matter (Trivedy and Goel, 1986; APHA, 1995; Aneja, 1996).

Primary Productivity

The primary productivity of phytoplankton was estimated by following the Gaarder and Gran's method (Gaarder and Gran, 1927). A set of light and dark BOD bottles were filled with surface water and closed tightly. The set of filled bottles were suspended about 10-15 cm below the surface of water using floats at their respective sites. The time of incubation of 3.00 hours from 10.00 a.m. to 1.00 p.m. was constantly maintained throughout the study period. To note the initial concentration of oxygen in the water, one more BOD bottle was filled separately before the incubation. Further, the initial and incubated samples were fixed immediately with modified Winkler's reagents at the sites of the study. Fixed BOD bottles were carried to the laboratory for further analysis. The modified Winkler's method was followed to note the dissolved oxygen of fixed samples of water to estimate the primary-productivity of the water body. The values of dissolved oxygen of initial and incubated light and dark bottles were converted to carbon value by multiplying with the factor of 0.375 to estimate the gross primary productivity (GPP), net primary productivity (NPP) and community respiration (CR) as follows:

$$\text{GPP: } gCm^{-3}hr^{-1} = \left(\frac{Dl-Dd}{hr}\right) \times 0.375$$

$$\text{NPP: } gCm^{-3}hr^{-1} = \left(\frac{Dl-Di}{hr}\right) \times 0.375$$

$$\text{CR: } gCm^{-3}hr^{-1} = \left(\frac{Di-Dd}{hr}\right) \times 0.375$$

Note:

Di: Dissolved oxygen in an initial bottle in mg/l

Dl: Dissolved oxygen in the light bottle in mg/l

Dd: Dissolved oxygen in the dark bottle in mg/l

hr: Duration of incubation in hours.

Results and Discussion

Water in nature is not pure and it contains dissolved salts, buffers, nutrients, etc. The exact concentrations are dependent on local conditions. The limno-chemical characteristics of Papnash pond and Bhalki tank have been analyzed and values are presented in Table 9.1.

Table 9.1: Limno-chemical Factors of Papnash Pond, Bidar and Bhalki Tank, Bhalki (February 2002 to January 2004)

Parameteres	*Papnash Pond*		*Bhalki Tank*	
	February 2002 to January 2003	*February 2003 to January 2004*	*February 2002 to January 2003*	*February 2003 to January 2004*
Atmospheric temperature	23.8–35	25–38.5	23.8–28.8	25–38.5
Water temperature	20.5–28.8	20.3–29.3	20.8–28.8	20.3–29.3
pH	7.1–7.85	7.2–8	7.1–7.9	7.2–7.8
Dissolved oxygen	4.84–12.3	7.7–11.54	4.9–12.3	7.7–11.5
Carbon dioxide	00–20.9	00–21.7	00–20.9	00–21.7
Alkalinity	110–215	115–150	110–215	115–150
Hardness	89–162	109–193	89–162	109–193
Calcium	18–31	18–51	18–31	18–51
Magnesium	9–21	8–21	9–21	8–21
Chlorine	19–49	23–72	19–49	24–72
Nitrite	0.003–0.377	0.012–0.448	35–89	42–130
Phosphate	0.003–0.288	0.012–0.069	0.003–0.377	0.012–0.448
Silicon	4.4–25.7	3.3–13.3	0.003–0.288	0.012–0.069
Sulphate	2.2–12.45	2.88–11.65	4.4–25.7	3.3–13.3
Total solids	13–161	53–205	2.2–12.5	2.9–11.7
Organic matter	0.32–2.65	0.24–3.02	13–161	53–205
GPP	51–477	70–380	26–522	102–405
NPP	–127–405	–177–155	–228–154	–278–228
CR	43–203	51–304	76–532	76–380

All the values are expressed in mgl^{-1} except pH, atmospheric temperature and water temperature. The study of primary production is concerned with evaluation of capacity of an ecosystem to build up, organic compounds of high chemical potentials for further transformation and flow to higher trophic levels. Photosynthesis and chemosynthesis are the fundamental processes of primary production. Primary production studies in aquatic environments are however restricted to or centered on the flow of carbon. Primary productivity of phytoplankton can be measured directly in contrast to the productivity of other trophic levels. The phytoplankton primary productivity in water bodies of Bidar district is presented in Table 9.2 and 9.3. The Pearson's correlation matrix between limno-chemical factors and primary productivity has been computed and presented in Table 9.4.

Table 9.2: Primary Productivity of Papnash Pond, Bidar (February 2002 to January 2004)

Months	*Site–A*			*Site–B*		
	Gross Primary Productivity	*Net Primary Productivity*	*Community Respiration*	*Gross Primary Productivity*	*Net Primary Productivity*	*Community Respiration*
February-02	245	195	051	051	-101	153
March	203	113	090	253	229	024
April	463	404	059	431	405	026
May	101	-051	151	203	051	153
June	051	-153	203	101	-101	203
July	153	103	051	304	253	051
August	253	101	151	101	051	051
September	303	253	051	101	000	101
October	051	-051	101	051	-051	101
November	051	-051	101	000	-051	051
December	253	200	051	176	101	074
January-03	101	051	051	051	-051	101
February	153	101	051	101	051	051
March	203	151	051	254	051	203
April	606	259	348	153	051	101
May	101	-153	254	051	-153	203
June	253	-051	304	000	-303	303
July	153	051	101	051	-101	153
August	051	-051	101	455	253	203
September	101	-051	153	153	051	101
October	051	-101	151	304	253	051
November	153	101	051	354	101	254
December	101	051	051	106	045	151
January-04	101	051	051	304	253	051

All the values are expressed as $gCm^{-3}hr^{-1}$

Table 9.3: Primary Productivity of Bhalki Tank, Bhalki (February 2002 to January 2004)

Months	*Site–A*			*Site–B*		
	Gross Primary Productivity	*Net Primary Productivity*	*Community Respiration*	*Gross Primary Productivity*	*Net Primary Productivity*	*Community Respiration*
February-02	304	203	101	203	051	153
March	253	-253	505	658	304	354
April	455	-094	549	250	151	101
May	203	051	151	455	051	405
June	558	-100	658	304	253	051
July	203	-455	658	405	000	405
August	304	253	051	151	051	101
September	608	308	300	455	000	455
October	355	254	101	101	000	101
November	419	196	223	101	000	101
December	051	000	051	253	051	203
January-03	051	000	051	000	-101	101
February	203	101	101	253	151	101
March	000	-354	354	203	-203	405
April	455	253	203	101	-051	153
May	153	101	051	253	051	203
June	203	101	101	606	051	556
July	455	354	101	153	101	051
August	151	101	051	101	-151	253
September	658	203	455	101	-151	253
October	151	101	051	354	253	101
November	101	051	051	203	101	101
December	203	051	153	000	-153	153
January-04	505	201	304	059	-043	101

All the values are expressed as $gCm^{-3}hr^{-1}$

Gross Primary Productivity

The gross primary productivity of Papnash pond values ranged up to 606 $gCm^{-3}hr^{-1}$. The complete absence of productivity was noticed in June 2003 and the highest productivity was recorded in April 2003 at site A (Table 9.2). The variation of GPP ranged up to 658 $gCm^{-3}hr^{-1}$ in Bhalki tank. The total absence of productivity was observed during March 2003 at site A and January and December 2003 at site B. The maximum productivity was recorded in September 2003 at site A and March 2002 at site B (Table 9.3).

The complete absence of GPP in Papnash pond in the beginning of monsoon of 2003 was due to low phytoplankton population together with other factors like dilution of nutrients, cloudy weather and low temperature. Similarly in Bhalki tank during the beginning of summer of 2003 and in post winter months of 2003 no blooms were observed. Such findings of complete absence of GPP during monsoon was also recorded by Rajkumar (2004) in KAK pond, Coimbatore, and in summer by Sukumaran (2002) in Lalbagh tank of Bangalore and in winter by Eshwaralal and Angadi (2002) in Jagat tank in Gulbarga.

Table 9.4: Pearson's Correlation Coefficients of Limno-chemical Factors and Primary Productivity of Papnash Pond, Bidar and Bhalki Tank, Bhalki (February 2002 to January 2004)

Limno-chemical	*Papnash Pond*			*Bhalki Tank*		
	Gross Primary Productivity	*Net Primary Productivity*	*Community Respiration*	*Gross Primary Productivity*	*Net Primary Productivity*	*Community Respiration*
Atmospheric temp.	+0.593**	+0.198	+0.089	+0.103	–0.007	+0.070
Water temp.	+0.120	–0.050	+0.219	+0.239	–0.028	+0.207
pH	+0.344	+0.142	+0.304	–0.024	–0.306	0.200
Diss. Oxygen	+0.419*	+0.182	+0.041	+0.349	–0.182	0.369
Carbon dioxide	–0.182	+0.282	–0.034	+0.387	+0.158	0.236
Alkalinity	+0.168	–0.130	+0.298	+0.101	–0.528**	0.505*
Hardness	–0.004	–0.390	+0.466**	+0.358	0.084	+0.237
Calcium	–0.188	–0.355	+0.204	0.282	0.372	–0.045
Magnesium	+0.025	–0.097	+0.375	+0.131	–0.080	+0.175
Chlorine	–0.043	–0.277	+0.305	–0.126	–0.190	+0.048
Nitrite	+0.068	–0.113	–0.078	+0.017	–0.370	+0.288
Phosphate	+0.030	–0.228	+0.466*	+0.134	–0.426*	0.450*
Silicon	–0.118	–0.485*	+0.690**	–0.094	+0.211	–0.220
Sulphate	–0.263	–0.337	0.228	–0.033	0.173	–0.164
Total solids	+0.137	–0.124	+0.384	+0.250	+0.303	–0.064
Organic matter	+0.114	–0.191	+0.332	+0.085	–0.261	+0.271
GPP	+1.000			+1.000		
NPP	+0.767**	+1.000		+0.386	+1.000	
CR	+0.239	–0.613**	+1.000	0.518**	–0.576	+1.000

* is significant at the 0.05 level (2-tailed); ** is significant at the 0.01 level (2-tailed).

The highest GPP of 606 $gCm^{-3}hr^{-1}$ and 658 $gCm^{-3}hr^{-1}$ was recorded during summer month of 2003 in Papnash pond and of 2002 in Bhalki tank. This is due to increase in water temperature and alkalinity. Increase of GPP during post monsoon month of 2003 in Bhalki tank was also observed and it is due to the nutrients and organic matter that gets loaded and are available for production. Similarly, the maximum GPP was recorded during Summer by Jeyachandran and Krishnan (2001) and Rajkumar (2004) and in Monsoon by Vijay Kumar *et al.* (2000) and Das (2002) in their study.

Gross primary productivity has shown a good relationship with temperature, dissolved oxygen in Papnash pond (Table 9.4). Similar observations were recorded by Prakasam and Joseph (2000) and Sukumaran (2002).

Net Primary Productivity

The NPP values in Papnash pond ranged between -303 $gCm^{-3}hr^{-1}$ and 404 $gCm^{-3}hr^{-1}$. The highest utilization of stored food was more than its production was in June 2003 at site B and the maximum storage was in April 2002 at site A (Table 9.2). NPP varied from -455 $gCm^{3}hr^{-1}$ to 354 $gCm^{3}hr^{-1}$ during the study period in Bhalki tank. The highest utilization of stored food was more than its production in July 2002 and the maximum storage in July 2003 at site A (Table 9.3).

Net primary production has shown negative values of -303 $gCm^{-3}hr^{-1}$ and -455 $gCm^{-3}hr^{-1}$ during monsoon month in Papnash pond and Bhalki tank respectively, in Papnash pond in 2003 and in Bhalki tank in 2002. It may be because of the oxidation of bloom or may be sedimentation or both in the system. This finding is in agreement with the findings of Jeyachandran and Krishnan (2001) and Mali and Gajaria (2004). The negative value of NPP is an indication of increase of microbial fauna in water body.

The maximum NPP of 404 $gCm^{-3}hr^{-1}$ and 354 $gCm^{-3}hr^{-1}$ observed respectively in Papnash pond and Bhalki tank. In Papnash pond during summer month of 2002, it is due to availability of nutrients and favorable condition for the setting of algal bloom and in Bhalki tank during monsoon month of 2003 is due to the runoff carried from the surrounding regions of the water body. Eshwaralal and Angadi (2002) have also reported the similar observation of increase of NPP during summer in Jagat tank, Gulbarga, and Mali and Gajaria (2004) have in fish culture pond, Gujarat. Similarly, during Monsoon by Govindsamy *et al.* (2000) have in Biotopes of Coromandel coastal zone.

NPP values were found to fluctuate with the values of GPP (Table 9.4). This indicates that, no interference by primary consumers like zooplankton and others in Papnash pond.In Bhalki tank, Ca content had a positive relation with the GPP. The limno-chemical factors, such as alkalinity, nitrite and phosphate have shown a negative relation with NPP and it is the indication of utilization of carbonates as a source of carbon. Phosphate is known to act as nutrient sources for the growth and production of phytoplankton. Similarly, Eshwaralal and Angadi (2002) have also recorded negative relation between phosphate and NPP in Jagat tank, Gulbarga. Negative relation of NPP with alkalinity was also recorded by Mahadev *et al.* (2003) in Dalvoi lake, Mysore.

Community Respiration

The variation of CR from 24 $gCm^{-3}hr^{-1}$ to 348 $gCm^{-3}hr^{-1}$ was noticed in Papnash pond. The minimum CR was recorded in March 2002 at site B and the maximum in April 2003 at site A (Table 9.2). The CR fluctuated between 51 $gCm^{-3}hr^{-1}$ to 658 $gCm^{-3}hr^{-1}$ in Bhalki tank. The lowest CR was observed in August and December 2002 and January, May, August, September and November 2003 at site A and in June 2002 and July 2003 at site B. The highest CR was noticed in June and July 2002 at site A (Table 9.3).

The lowest CR of 24 $gCm^{-3}hr^{-1}$ and 51 $gCm^{-3}hr^{-1}$ was recorded respectively in Papnash pond and Bhalki tank. In Papnash pond during the early month of summer of 2002 and in Bhalki tank during all the seasons through out the study period. This may be due to influx of runoff from the surrounding fields and from the city that disturbs the microbial fauna in the water body. Ravindra Singh (1990) and Sukumaran (2002) have also recorded similar low CR values during summer in North and South tanks of Jamalpur and in Lalbagh tank, Bangalore respectively.

The maximum CR of 348 $gCm^{-3}hr^{-1}$ and 658 $gCm^{-3}hr^{-1}$ was recorded respectively in Papnash pond and Bhalki tank. In Papnash pond, it was during summer month of 2003 and it is because of higher phytoplankton and zooplankton population. In Bhalki tank, it was during monsoon month of

2002 and it may be due to enrichment of nutrients from surrounding fields. It increased the productivity and thereby it supplies oxygen for the metabolic activity of grazing zooplanktons in the water body. Ravindra Singh (1990) and Vijay Kumar *et al.* (2000) have also recorded high CR during summer in Lalbagh tank, Bangalore and Gobbur tank, Gulbarga and in monsoon by Bais *et al.* (1997) in Sagar lake, Sagar.

Community respiration has increased with the increase of phosphate in both water bodies and with hardness and organic matter in Papnash pond and alkalinity in Bhalki tank (Table 9.4). A negative relation of CR with NPP in both water bodies was observed. These finding are in agreement with the findings of Singh and Singh (1999) and Eshwaralal and Angadi (2002) have worked on ponds of Patna and Jagat tank at Gulbarga respectively.

Conclusion

The gross primary productivity ranged up to 606 $gCm^{-3}hr^{-1}$ in Papnash pond and 658 $gCm^{-3}hr^{-1}$ in Bhalki tank. The GPP was positively correlated with atmospheric temperature and dissolved oxygen in Papnash pond. The net primary productivity ranged between–303 $gCm^{-3}hr^{-1}$ and 404 $gCm^{-3}hr^{-1}$ in Papnash pond and -455 $gCm^{-3}hr^{-1}$ and 354 $gCm^{-3}hr^{-1}$ in Bhalki tank. NPP positively fluctuated with GPP in Papnash pond and negatively with phosphate in Bhalki tank. The variation of community respiration from 24g $Cm^{-3}hr^{-1}$ to 348 $gCm^{-3}hr^{-1}$ in Papnash pond and 51 $gCm^{-3}hr^{-1}$ to 658 $gCm^{-3}hr^{-1}$ in Bhalki tank is recorded. The CR fluctuated positively with hardness, phosphate and organic matter and negatively with NPP in Papnash pond.In Bhalki tank, CR has shown positive relation with alkalinity and phosphate and negative relation with NPP in Bhalki tank.

The present study clearly indicates that, water bodies are very efficient manufacturers of the fundamental organic matter. It also indicates that, there are several environmental factors acting simultaneously in the production capacity of the water body. Such as local hydrography, seasonal variability and limno-chemical state are factors, involved in the occurrence of dissolving nutrients in both water bodies. Papnash pond water body possesses relatively more nutrients than Bhalki tank. However, the influences of parameters, which bring about fluctuations in primary production levels, are given much importance. Limno-chemical factors like atmospheric temperature, dissolved oxygen, hardness, phosphate and organic matter in Papnash pond and phosphate and alkalinity in Bhalki tank were mainly responsible to regulate the phytoplankton production. Results are also throw light on that, the water characteristics of Papnash pond were favorable for productivity than the Bhalki tank. Attention should be paid immediately to Papnash pond, failing which there will be concomitant loss of valuable water resource due to high eutrophication.

Acknowledgements

The authors are indebted to late Dr. S.B. Angadi, Reader, Dept. of Botany, for his valuable guidance. We also thank the head of the Dept. of Botany, Gulbarga University, Gulbarga; Assistant Director I/C, RRI, Bangalore and Director, CCRAS, New Delhi for providing the facilities and encouragement.

References

Aneja, K.R., 1996. *Experiments, in Microbiology, Plant Pathology, Tissue Culture and Mushroom Cultivation, 2nd edn.* Wishwa Prakashan, New Delhi.

Angadi, S.B., Lingannaiah, B. and Eshwarlal, S., 1999. Limnological studies of Jagat tank, Gulbarga. In: *Freshwater Ecosystem of India.* Daya Pub. New Delhi, p. 133–159.

APHA, AWWA, WPCF, 1995. *'Standard Methods for the Estimation of Water and Wastewater'*. American Public Health Association, American waste water Association and Water Pollution Control Federation 19th Edition Washington D.C.

Ayyappan, S. and Gupta, T.R.C., 1985. Limnology of Ramasamudra tank. Primary production. *Bull. Bot. Soc. Sagar*, 32: 82–88.

Bais, V.S., Agrawal, N.C. and Tazeen, A., 1997. Seasonal changes in phytoplankton productivity due to artificial enrichment of nutrients. A situ experiment. *J. Environ. Biol.* 18(3): 249–255.

Chacko, P.M. and Krishnamurthy, P.I., 1954. On the plankton of three fresh water Fish ponds in Madras city, India. *Indo Pacific Fish. Coun.*, (UNESCO), p. 103–107.

Das, A.K., 2002. Phytoplankton primary production in some selected reservoirs of Andra Pradesh. *Geobios.*, 29: 52–57.

Eshwaralal, S. and Angadi, S.B., 2002. Primary productivity of two freshwater bodies of Gulbarga, India. *Nat. Environ. and Poll. Tech.*, 1(2): 151–157.

Gaarder, T. and Gran, H.H., 1927. Investigations of the production of plankton in the Oscofiord. *P.V.R. Rapp. Reun. Commn. Int. Explor. Scient. Mer. Meditter.*, 42: 1–48.

Gaddad, S.M., Nimbargi, P.M. and Rodgi, S.S., 1983. Ecological studies of two polluted water bodies of Gulbarga. *Poll. Res.*, 2(2): 49–52.

Ganapati, S.V., 1943. An ecological studies of a garden pond containing abundant Zooplankton. *Proc. Indian. Acad. Sci.*, 17: 41–58.

George, M.G., 1962. Occurrence of permanent algal bloom in a Fish tank at Delhi with special reference to factors responsible for its production. *Proc. Indian Acad. Sci.*, 56 (8): 354–362.

Gopal, B., Sharma, K.P. and Trivedy, R.K., 1978. Studies on ecology and production in Indian fresh water ecosystems at primary producer level with emphasis on macrophytes. In: *Glimpses of Ecology*, (Eds.) J.S. Singh and B. Gopal. Int. Sci. Pub., Jaipur, p. 349–376.

Govindasamy, C., Kannan, L. and Azariah, J., 2000. Seasonal variation in physico-chemical properties and primary production in the coastal water biotopes of Coromandel coast, India. *J. Environ. Biol.*, 21(1): 1–7.

Janakiram, K. and Jayaraj, Y.M., 1996. Effect of organophosphorous insecticides on primary productivity of stabilisation pond. *Geobios*, 23: 219–222.

Jeyachandran, A. and Krishnan, N., 2001. Hydrobiology and primary productivity of Viraganur dam, Madurai, Tamil Nadu. *Indian Hydrobiol.*, 4(1): 39–42.

Khan, A.A. and Siddiqui, A.R., 1971. Primary production in a tropical fish pond at Aligarh, India. *Hydrobiol.*, 37: 447–456.

Mahadev, J., Nagarathnamma and Hosmani, S.P., 2003. The significance of cluster analysis in determining relationship between physico–chemical parameters and phytoplankton in polluted waters. *J. Ecol.Toxicol. Environ. Monit.*, 13(1): 53–58.

Mali, K. N. and Gajaria, S.C., 2004. Assessment of primary productivity and hydrobiological characterization of a Fish culture pond, Gujarat. *Indian Hydrobiol.*, 7(1 and 2): 113–119.

Naganandini, M.N. and Hosmani, S.P., 1998. Ecology of certain inland waters of Mysore district occurrence of cyanophycean bloom at Hosakere lake. *Poll. Res.*, 17(2): 123–125.

Palaniappan, N., Marimuthu, R. and Rajendran, M., 1981. Seasonal variation in primary productivity of two rock pools in Salem, Tamil Nadu. *Papers on Freshwater Biology*, The Salem institute of Experimental Biology, Salem (TN), p. 58–72.

Prakasam, V.R. and Joseph, M.L., 2000. Water quality of Sasthamcotta lake, Kerala, (India) in relation to primary productivity and pollution from anthropogenic sources. *J. Environ. Biol.*, 21(4): 305–307.

Rajkumar, N., 2004. A review on the quantum of phytoplasktonic primary production of polluted fresh water pond. *Indian Hydrobiol.*, 7(1 and 2): 61–65.

Seenayya, G. and Zafar, A.R., 1979. An ecological study of the Mir Alam lake, Hyderabad, India. *Ind. J. Bot.*, 2(2): 214–220.

Singh, A.K. and Singh, D.K., 1999: A comparative study on the phytoplanktonic primary production of river Ganga and pond of Patna (Bihar), India. *J. Environ. Biol.*, 20(3): 263–270.

Singh, H.P., 1998. Studies on primary production in Gobindsagar reservoir, Himachal Pradesh. *J. Environ. Biol.*, 19(2): 167–170.

Singh, Ravindra, 1990. Correlation between certain physico-chemical parameters and primary production of phytoplankton at Jamalpur, Munger. *Geobios*, 17: 229–234.

Sreenivasan, A., 1964. Limnological features and primary production in a polluted moat at Vellore, Madras state. *Envi. Health.*, 6: 237–245.

Sreenivasan, A., 1969. The hydrobiological features, primary production and fisheries of stanley reservoir, Mettur dam for the year 1958 to 1965. *Madras J.Fish*, 5: 1–17.

Sukumaran, P.K., 2002. Primary production dynamics of a Perennial tank in Bangalore, Karnataka. *Geobios*, 29(1): 41–46.

Sukumaran, P.K., 2003. Diel variations in the biotic and abiotic parameters of a fresh water tank in Bangalore. *Geobios*, 30(4): 277–283.

Trivedy, R.K.and Goel, P.K., 1986. *Chemical and Biological Methods for Water Pollution Studies.* Environmental Publication, Karad, India.

Valecha, V. and Bhatnagar, G.P., 1989. Primary productivity of phytoplankton in a eutrophic lower lake Bhopal. India. *Envi.* and *Ecol.*, 7(1): 202–205.

Vijay Kumar, K. and Paul, R., 1992. Primary productivity of two standing water bodies of Dharwad, Karnataka. *Environ and Ecol.*, 10(2): 261–264.

Vijay Kumar, K. and Paul, R. 1994. Effect of nutrient enrichment on phytoplankton production. *Proc. Acad. Environ. and Biol.* 3 (2): 171–175.

Vijay Kumar, K., Anjinappa, H., Virupanna and Padmavathi, C., 2000. Seasonal variations in the primary productivity of Gobbur tank, Gulbarga. *Proc. Acad. Environ. Biol.*, 9(2): 119–123.

Chapter 10

Studies on Certain Physico-chemical Parameters of Hataikheda Reservoir Near Bhopal, India

Vikas Salgotra, S.A. Mastan, Adarsh Kumar and T.A. Qureshi

Introduction

Reservoirs are constructed by impounding the river run-off. They are vitally important in central India because of the role they play in generating hydro-electricity, irrigation and fish production and as most important source of the drinking water. Bhopal 'the city of lakes' is comparatively hot during the most part of the year, there are number of lakes and reservoirs except other small standing water bodies. Water in an important substance required by all living organisms from microorganisms to man and for all anthropogenic activities (Khatri, 1985; Dagaonkar and Saksena, 1992).

Hataikheda reservoir is a minor reservoir constructed at about 10 km away from Bhopal city near BHEL Township. On its south lies the inhabited area of Anand Nagar, the eastern and western sides of this water body is surrounded by agricultural fields while its northern side is rocky. It is situated at 32°16′ N latitude and 77°30′ longitude above 460 MSL. Its gross command area is 1148 acres and cultivable command area is 915 acres. The source of water for this reservoir is Ghoropachhar nullah a tributary of Betwa River. It serves as a source of irrigation to three villages and also supplies water for industrial use. It receives domestic as well as industrial wastes from the surrounding areas. The present work is designed to study its various abiotic factors.

Materials and Methods

The work was carried out during March 2003 to February 2004. Water and soil samples were collected from the lentic zone. Dissolved oxygen, water temperature, pH, Secchi disk transparency, Total dissolved solids, Free CO_2, Total alkalinity, Total hardness, Silicate, Calcium and Magnesium were analyzed by following standard methods (APHA 1995). Soil was analyzed for pH, available nitrogen, available phosphorous, carbonate and organic carbon following Trivedy and Goel (1984).

Results and Discussion

Physico-chemical Features of Water

The results obtained during the present investigation are depicted in Table 10.1.The temperature was found to be in the range of 18°C–33°C. The minimum water temperature was recorded in the month of December while the maximum was recorded in the month of May and June 2003.

The hydrogen ion concentration (pH) is moderate to high *i.e.,* 5-7.5. pH below 6.5 is not considered good for aquatic life but this low concentration is present only during the months of April to May 2003. While during the rest of study period pH remains in good ranges. According to Swingle (1961), reproduction and growth of fish diminishes if water pH persists below 6.5 and above 9.5 for a prolonged period.

The water transparency depends on the microorganisms, organic matter and suspended clay particles present in H_2O. Water transparency was moderate, 49.3- 58.3 cm. Maximum transparency was observed during pre-monsoon period. Sakhare and Joshi (2003) observed the transparency values 14–35 cm from Papnas wetland in Osmanabad district of Maharashtra.

Total dissolved solids ranged from 197–263 mg/l, the excessive total dissolved solids generally affects the palatability. In river the total dissolved solids is attributed to pollution bye effluents (Rajalakshmi and Sreelatha, 2005). Here also the high TDS value might be due to the wastes coming from the industries and other domestic waste from the inhabitant areas of Anand Nagar.

Table 10.4: Range of Physico-chemical Parameters of Hataikheda Reservoir Near Bhopal

Sl.No.	*Parameters*	*Range*
1.	Water temperature (°C)	18–33
2.	pH	5–7.5
3.	Transparency (cm)	49.3–58.3
4.	Total dissolved solids (mg/l)	197–263
5.	Calcium (mg/l)	17.8–29.3
6.	Magnesium (mg/l)	5.8–12.1
7.	Silicates (mg/l)	0–3.9
8.	Total hardness (mg/l)	108.1–158
9.	Dissolved oxygen (mg/l)	3.9–7.4
10.	Free carbon dioxide (mg/l)	4.5–6.4
11.	Total alkalinity (mg/l)	24.5–65.7

Calcium is one of the alkaline earth metals, the other being magnesium, barium, strontium etc. the calcium is not known to produce hazardous effects on human health. Its concentration is very significant and is ranged from 17.8 to 29.3 mg/l. Kulkarni's Rajendra Rao *et al.* (2002) observed the calcium level varied from 56.9–101.7 mg /l. Magnesium has ten times the solubility of calcium and being bivalent, it also make water hard its concentration ranged from 5.8 to 12.1mg/l.

The concentration of silicate just do not compensate with the normal value of silicate and is present in the range of 0-3.9 mg/l and is absent in the months of Oct., Nov. and Dec. During the monsoon months the concentration declines and in the later months the concentration seems to improve.

Total hardness ranges from 108.1mg/l to 158 mg/l. Total hardness is not as much important as Total alkalinity with regards to fish production.

Significant seasonal variations have been observed in the dissolved oxygen varied from 3.9-7.4 mg/l. This level is quite good in its ranges and is in adherent with the natural wind action which forms a ultimate environment with regards to the culture of various economically important aquatic animals. DO along with turbidity gives information about the nature of the aquatic ecosystem better than any other chemical parameters (Hutchinson, 1957). It has been observed that DO level more than 5 mg/l is good for the even growth of flora and fauna (Das, 2000).

Carbon dioxide is highly soluble in water; it is a minor constituent of the atmosphere and remains in equilibrium concentration giving acidic reactions in aquatic environment. Free carbon dioxide level varies from 4.5-6.4 mg/l, which just compensate with fisheries point of views. Due to the interdependence of pH with the carbonate system, the reduction CO_2 level as a result of photosynthesis by aquatic plants may also alter the pH of water (Pandey *et al.*, 2001).

Total alkalinity or acidic combining capacity of water is another important factor, which can be correlated with productivity and is generally caused by the carbonates and bicarbonates of calcium and magnesium. The total alkalinity varies from 24.5-65.7 mg/l. Moyle (1949); based on the study of large no. of lakes and ponds showed that alkalinity ranging from 40-90 mg/l gives medium productivity.

Physico-chemical Properties of Soil

The physico-chemical properties of soil are represented in Table 10.2.The soil is found to be compensating with the alkaline nature in good part of the year as pH ranges from 6.4-7.1. The available nitrogen ranges from 10.5-15 mg/100g and available Phosphorous is in the ranges 1- 4.31 mg/100g. The effect of such a low ranges if found to have restricted the plankton production (David *et al.*, 1969). The organic carbon was moderate, ranging from 1.9-3.5 per cent and carbonate between 1.5-11 per cent. The water quality remains more or less the same around the year. The soil has very little impact on water quality as water derives most of its nutrients from the catchment areas possibly.

Table 10.2: Physico-chemical Properties of Soil

Sl.No.	*Parameters*	*Range*
1.	pH	6.4-7.1
2.	Available Nitrogen (mg/100g)	10.5-15
3.	Available phosphrous (mg/100g)	1-4.31
4.	Organic carbon (per cent)	1.9-3.5
5.	Carbonate (per cent)	1.5-11

Acknowledgements

We wish to thank Dr. K. Borana, Mr. Satish Patil, Department of Applied Aquaculture, B.U. Bhopal, for suggestions and guidance.

References

APHA, 1995. *Standard Methods for Examination of Water and Wastewater Analysis*, 19th Edition. American Public Health Association, Washington, D.C.

Dagaonkar, A. and D.N. Saksena, 1992. Physico-chemical characteristics of temple tank, Katia Sagar, Gwalior. *Hydrobiol.*, 8: 18–19.

David, A.P., Roy, B.V., Govind, K., Roygopal, V. and Banerjee, R.K., 1969. Limnology and fishes of Tungabadhra reservoir. *Bull. Cent. Inland Fish. Res. Inst.*, Barrackpore, November, 13: 188.

Hutchinson, G.E., 1957. *A Treatise on Limnology*, Vol 1, Part 2: Chemistry of Lake. John Wiley and Sons, Inc., U.S.A.

Khatri, T.C., 1985. Physico-chemical feature of Idukkwi reservoir, Kerela during pre-monsoon period. *Environ Ecol.*, 3: 134–137.

Kulkarni Rajendra Rao, Sharma, R.N. and Burkari, M., 2002. Diurnal variations of physico-chemical aspects of pollution in Khushavati river at Quepem, Goa. *J. Aqua. Biol.*17(1):27–28.

Mazher Sultan and Dawood Sharief, 2004. Water pollution studies in the Double lake (Erretal eri) with relation to Phytoplankton. *J. Aqua. Biol.*, 19(1): 15–18.

Pandey, G.C., Neraliya, S., Srivastava, P.K. and Tripathi, G., 2001. Bioremediation: A biological tool for pollution abatement. In: *Current Topics in Environmental Sciences*, (Eds.) Tripathi, G. and Pandey G.C. ABD Publishers, Jaipur, India, pp. 347–366.

Rajalakashmi, S. and K. Sreelatha, 2005. Studies on the physicochemical parameters of river Gautami Godavari Yanam (Union Territory of Pondicherry). *J. Aqua. Biol.*, 20(2): 110–112.

Sakhare, V.B. and Joshi, P.K., 2003. Physico-chemical limnology of Papnas: A minor wetland in Tuljapur town, Maharashtra. *J. Aqua. Biol.*, 18(2): 93–95.

Swingle, H.S., 1961. Relationship of pH of pond water to their suitability for fish culture. *Pacific Sci. Congress*, 9(1957), 10: 72–75.

Trivedy, R.K. and Goel, P.K., 1984. *Chemical and Biological Methods for Water Pollution Studies*. Environmental Publication, Karad, India.

Chapter 11

Water Quality of Bastawade Pond from Tasgaon Tahsil of Sangli District (Maharashtra) and its Significance to Prawn Farming

S.A. Khabade and M.B. Mule

Introduction

Water quality plays an important role in the growth of aquatic animals. There should be optimum range of water quality standards for the normal growth of aquatic animals.The water quality standards below and above the optimum range may lead to either stress or death among the aquatic animals.Physico-chemical and microbial characteristics of water body depend upon many factors such as location, dispersal of domestic and industrial waste water and many other activities (Sathe *et al.*, 2000). Various water quality standards are recommended by World Health Organization (WHO), Indian Council for Medical Research (ICMR), Indian Standard Institute (IS) and National Environmental Engineering Research Institute (NEERI).

For the normal growth of the freshwater prawns, the optimum range of water quality is also essential. There are about 100 species of fresh water prawn *Macrobrachium*, identified from various parts of the world of which about 25 species are found in India. Among the Indian prawns, *Macrobrachium rosenbergii, M. malcolmsonii, M. brimanicum, M. idella, M. rude, M. equidens, M. seacriculum* and *M. lamarrei* are suitable for farming in India. In India, freshwater prawn farming is a growing industry. Among the different species of freshwater prawn, the giant fresh water prawn, *Macrobrachium rosenbergii* (also known as 'Scampi' in the trade) is the most commercially important species because of its ability to grow both in fresh water and low saline waters, compatibility for polyculture with carps, omnivorous feeding habits, hardihood and potential to grow fast (Dholakia, 2004). It is seen in almost all major rivers and lakes in India, which are connected to sea. It does not have much of problems of different disease attacks and has good consumer preference. It has ever increasing demand both in the domestic as well as export market. Prawn farming is a profitable venture in terms of export value.

Gaint freshwater prawn is a native of India, so that it makes the farming operation easier. In India, there are about 18 million acres of freshwater area suitable for freshwater prawn farming. Mostly, prawn can be grown with other species of fish which is known as polyculture. Prawn farming in ponds is the most technically and economically viable farming method. Eventhough fresh water prawn production from unit are in unit time is less compared to fish, the high market price for prawns makes prawn farming more attractive than fish farming.

The fishery department of Sangli district has grown the prawns in Bastawade shallow water pond. As there is no sufficient baseline data about water quality parameters of the pond under study, hence the present work was undertaken to monitor the water quality parameters, which helps to decide suitability of water for prawn farming.

Study Area

The Bastawade shallow water pond is located in Tasgaon tahsil of Sangli district in Maharashtra. The pond is situated in the eastern draught prone hilly area of Tasgaon tahsil, in between Lat. 16°46′ to 17°1′ N and Log 73°42′ to 75°4′ E.

Materials and Methods

Monthly collection of water sample was done from the period July 2000 to April 2001 by using samplers (Plastic containers of 5 liter size). The site selected for water sampling lies near the pond's earthen embankment.

The physico-chemical parameters such as water temperature, pH and dissolved oxygen were determined in the field during monthly visits. The total alkalinity, acidity, hardness, magnesium, calcium, chlorides, residual chlorine, free CO_2, hydrogen sulphide, nitrate and phosphate were analysed in laboratory by preserving the water with preservative. The analysis of physico-chemical parameters was done according to the standard method of Trivedy and Goel (1984), APHA, AWWA and WPCF (1985).

Results and Discussion

There is no substitute for good water quality and quantity for fresh water prawn farming. The physico-chemical nature of water and its seasonal variation should be studied at the time of site selection for prawn farming.

Fourteen water quality parameters of 'Bastawade shallow water pond' were determined regularly in each month and results are summarized in the Table 11.1. During months of June 2000 and May 2001 the pond was dry.

The physico-chemical characteristics of water recorded are affected by rainfall, temperature, availability of light.

In present investigation the water temperature ranged between 25°C to 34°C and shows seasonal variation. Basha and Qureshi (1993) observed that the temperature ranging between 24-28°C was conductive for the abundant occurrence of prawns. The optimum range of temperature for best growth of prawn is about 28°C to 31°C (MPEDA 1994). Gopinathan *et al.* (1978) have reported direct relationship between temperature and number of juveniles of brackish water prawns.

Table 11.1: Water Quality Parameters of 'Bastawade Shallow Water Pond' from July 2000 to April 2001

Parameters	*Months*									
	Jul 2000	*Aug 2000*	*Sept 2000*	*Oct 2000*	*Nov 2000*	*Dec 2000*	*Jan 2001*	*Feb 2001*	*March 2001*	*April 2001*
Water temperature (°C)	26.0	27.0	28.0	27.6	26.0	25.0	27.0	28.0	31.0	34.0
pH	7.8	8.0	8.2	8.5	8.4	8.3	7.7	7.2	7.1	7.4
Total alkalinity (mg/l)	220	265	150	195	220	220	245	195	200	155
Acidity (mg/l)	42.5	60.0	22.5	20.0	32.5	37.5	30.0	27.5	30.0	20.0
Hardness (mg/l)	216	202	126	122	110	134	140	184	116	200
Magnesium (mg/l)	93.62	94.94	13.18	35.60	5.27	39.55	35.60	14.50	10.54	77.79
Calcium (mg/l)	29.65	23.24	30.46	27.25	51.30	29.65	34.46	24.84	56.11	32.86
Chlorides (mg/l)	42.6	53.96	32.66	42.6	48.28	39.76	42.6	32.66	34.08	137.74
Residual Chlorine (mg/l)	Ab	Ab	Ab	Ab	Ab	Ab	Ab	Ab	Ab	Ab
Dissolved Oxygen (mg/l)	2.41	3.82	4.63	7.04	6.64	7.65	7.44	8.65	5.83	4.02
Free CO_2 (mg/l)	17.6	11.0	Ab	Ab	Ab	Ab	Ab	31.0	19.8	11.0
Hydrogen Sulphide (mg/l)	0.14	0.14	0.28	0.425	0.85	0.425	0.566	0.708	3.11	0.85
Nitrate (mg/l)	3.9	4.3	14.4	2.3	0.8	0.3	1.3	1.6	1.8	4.6
Phosphate (mg/l)	0.08	0.01	1.0	0.3	0.01	1.2	0.6	0.6	2.6	2.0

All values are mean of four readings; All values expressed in mg/L except water temperature and water pH.

pH depends on quality of water, photosynthesis, level of ammonia, metabolic wastes etc. The pH of Bastawade pond was alkaline and values ranged from 7.1 to 8.5, being lower in summer and higher in winters. Higher the pH, higher will be the chances for ammonia hazard. In the present study the pH was always found on alkaline side which is considered favourable for aquatic animals like prawns. The lower pH values in summer are observed as a result of decomposition of organic matter at higher temperature, resulting into surplus amount of free carbondioxide (Saha and Pandit, 1986).

Total alkalinity of Bastawade pond ranged between 150 to 265 mg/l. The maximum values were recorded in the month of August 2000 due to less rainfall and minimum values were recorded in the month of September 2000, due to high rainfall. The alkalinity being due to bicarbonates (Misra *et al.*, 1975 and 1976; Bohra, 1976; Bohra *et al.*, 1979). The accumulation of bicarbonate in summer may be due to increased rate of decomposition (Munawar 1970). Basha and Qureshi (1993) have also reported the significant direct correlation of prawn population with alkalinity and total hardness.

Acidity in natural unpolluted freshwaters is mostly due to the presence of free CO_2 in the form of carbonic acid, Trivedy and Goel (1984). In Bastawade pond the acidity reported was ranged between 20.0 to 60.0 mg/l. The abundant occurrence of prawns matched with the peak of alkalinity and total hardness (S. Manohar and Qureshi 1994).

Hardness is generally caused by the calcium and magnesium ions present in water. The values of hardness ranged between 110 to 216 mg/l. Hardness between 50 to 100 mg/l is permissible for prawn growth and it depends on the basic source of water (MPEDA 1994). Liming raises total hardness.

Magnesium and Calcium have been shown to be increased by sewage pollution (Beeton, 1969; Munawar, 1979). In Bastawade pond the minimum amount of magnesium found was 5.27 mg/l, reported during November 2000. The higher amount of magnesium found was 94.94 mg/l reported during August 2000. The minimum amount of calcium found in Bastawade pond was 23.24 mg/l and the maximum amount reported was 56.11 mg/l.

Chlorides show a range of 32.66 to 137.74 mg/l. According to Thresh *et al.* (1944), chlorides are generally considered as indicator of pollution. The prawn population shows inverse relation with chloride (Manohar and Qureshi, 1994). Similar observations were also reported by Bash and Qureshi (1993).

The residual chlorine was absent throughout the year of investigation in Bastawade shallow water pond.

The dissolved oxygen is stated to be directly related with the prawn abundance (Broad 1962; Basha and Qureshi 1993). According to Manohar and Qureshi (1994), the highest peak of dissolved oxygen found in winter fall coincided with the peak of prawn abundance. However, during the pre-monsoon there is peak of prawn abundance and the dissolved oxygen has been found at moderate level. Exactly, the same results were obtained about dissolved oxygen in Bastawade pond. The peak values of dissolved oxygen were recorded in winter are 7.65, 7.44 and 8.65 mg/l.However, during pre-monsoon months the dissolved oxygen reported was found to be at moderate level, about 5.83 and 4.02 mg/l.

Qureshi (1994) stated that the occurrence of prawns generally co-incided with the lowest ebb of free carbon dioxide, particularly winter when the free carbon dioxide was found completely absent. However, the pre-monsoon peak of the prawns is observed when the free carbondioxide was present. In the present investigation also, free CO_2 was absent during winter and present during pre-monsoon period and during rainy period. The most of the growth of prawns in Bastawade pond takes place during winter and later few months of summer.

In water reservoirs all the sulphur from organic matter is released in the form of H_2S and later, may get converted into elemental sulphur or sulphate depending upon pH and oxygen regime. Sulphur can be considered as an indicator of organic pollution and prevailing oxygen deficient conditions (Ruttener 1953 and Vamos, 1964).

Presence of H_2S in water body is an indication of organic pollution load in it (Olsen and Sommerfeld, 1977). In the present investigation the concentration of H_2S in the Bastawade pond is not much higher which indicate that there is no high level of organic pollution. The H_2S range found was 0.14 mg/L to 3.11 mg/L which is not harmful for prawn growth.

The nitrates remain much variable throughout the year. According to Ganapati (1960), the non-polluted tropical waters are generally deficient in nitrate but the factors like discharge of sewage, surface runoff and nitrogen fixation may increase its concentration in the water body. In present investigation it has been observed that whatever nitrate present during the year of investigation is due to the surface runoff and nitrogen fixation and not due to the discharge of sewage. Nitrate ranged from 0.3 mg/l to 14.4 mg/l.

The values of phosphates in the present investigation ranged between 0.01 mg/l to 2.6 mg/l. High values of phosphates usually found in reservoirs affecting by domestic sewage (Anantharaj *et al.*, 1987). In Bastawade pond there is no any type of domestic sewage pollution, so that the values of phosphates recorded were not higher but in optimum range. High phosphate may lead to significant, undesirable growth of plankton.

The optimum range of nitrates and phosphates recorded not affecting the growth of the prawns in Bastawade pond.

In the present study, the prawn species *Macrobrachium rosenbergii* which is cultivated in the man made water reservoir, the Bastawade pond shows the luxuriant growth rate. The average weight of the prawns recorded was 100 to 130 gms in 6 to 7 months. The standard period of good growth of prawns is about 8 months to 12 months. Within 5 months the prawns grows upto 50 to 100 gms in weight.

In Bastawade shallow water pond, the post larvae of prawns were deposited during October 2000 and collected the well grown prawns during April 2001. Thus within 7 months, there is a good growth of the prawns, *M. rosenbergii.* It indicates that the water quality (physico-chemical and biological) of Bastawade shallow water pond is good for the growth of the prawns and for fresh water prawn farming in future. According to Saumen Chakrabarti (2005), abundance of *Macrobrachium dayanum* depended upon ecological balance between different biotic and abiotic (physico-chemical) factors. According to Qureshi (1994) the lower abundance of prawns in the vegetation free area was due to the sunlight factor. The water reservoir under study was vegetation free and oligotrophic in nature.

References

Ananthraj, B.V., Bhagyalakshmi and R. Lakshmi, 1987. Limnology of river Cooum with special reference to sewage and heavy metal pollution. *Proc. Indian Acad. Sci. (Anim. Sci.),* 96: 141–149.

APHA, AWWA, WPCF, 1985. *Standard Methods for the Examination of Water and Wastewater,* 20th Ed. American Public Health Association, Washington DC.

Basha, S.M. and Qureshi T.A., 1993. Hydrobiological characteristics and prawn population of Halali reservoir of Madhya Pradesh, India. *J. Applied Pure Biol.,* 8: 55–62.

Beeton, A.M., 1969. Changes in the environment and biota of the great lakes. In: *Eutrophication: Causes, Consequences and Correctives.* Nat. Acad. Sci., Washington, DC, pp. 150–187.

Bohra, O.P., 1976. Some aspects of limnology of Padamsagar and Ranisagar, Jodhpur. *Ph.D. Thesis,* Jodhpur University, Jodhpur.

Bohra, O.P., Misra, S.D. and Bhargava S.C., 1979. Diurnal variation studies on Nakhi lake, Mount Abu, India. *Bioresearch,* 3: 33–43.

Broad, A.C., 1962. Environmental requirement of Shrimp. *Publs. Publi. Hlth. Serv.* Washington, 1.

Chakrabarti, Saumen, 2005. Influence of biotic and abiotic factors on the occurrence and abundance of *Macrobachium dayanum.*

Dholakia, A.D., 2004. *Fisheries and Aquatic Resources of India.* Daya Publishing House, Delhi, p. 185.

Ganapati, S.V., 1960. Ecology of tropical waters. In: *Proceedings of Symposium on Algology,* (Ed.) D. Raghavan and P. Kachrro. ICAR, New Delhi, pp. 204–218.

Gopinathan, K., Kaliyanmurthy, M. and Rao, J., 1978. Studies on some species of post-larval peneids of lake Pulicat in relation to their environmental parameters. *Proc. Nat. Acad. Sci. Ind.,* 14: 195–209.

Manohar, S. and T.A. Qureshi, 1994. Populations of prawns in relation to the ecological characteristics of river Betwa at Bhojpur, Madhya Pradesh, India. Palani Paramount Publications ISSN: 0970–937–0006–241.

Misra, S.D., Bhargava, S.C. and Bohra, O.P., 1975. Diurnal variation in physico-chemical factors at Padamsagar reservoir during pre-monsoon period of the year 1974. *Geobios,* 2: 32–33.

MPEDA, 1994. *Fresh Water Prawn Farming.*

Munawar, M., 1979. Limnological studies on freshwater ponds of Hyderabad, India II : The biocenose. Distribution of unicellular and colonial phytoplankton in polluted and unpolluted environments. *Hydrobiol.*, 36(1): 105–128.

Olsen, R.D. and Sommerfeld, M.R., 1977. The physico-chemical limnology of a desert reservoir. *Hydrobiol.*, 55: 117–129.

Qureshi, T.A., 1994 Results of studies on freshwater prawn carried out in Madhya Pradesh. AFSIB. Special Publ. No 10 Mangalore, pp. 43–48.

Ruttener, F., 1953. *Fundamentals of Limnology*. University of Toranto Press.

Saha, L.C. and Pandit, B., 1986. Lentic and Lotic systems a review. *Acta Ecologica*, 8: 1–12.

Sathe, Sanjay S., Khabade, S.A.and Hujare M.S., 2000. Studies on wetlands of Tasgaon tahsil and its importance in relation to fisheries and agricultural productivity, Project Report submitted to University Grants Commission, New Delhi, Western Regional Office, Pune.

Thresh, J.C., E.V. Suckling and J.E. Beale, 1944. *The Examination of Water and Water Supplies*. London.

Trivedy, R.K. and P.K. Goel, 1984. *Practical Methods in Ecology and Environmental Science*. Enviro Media Publications, Karad (India).

Vamos, R., 1964. The release of hydrogen sulphide from mud I. *Soil. Sci.*, 15(1): 103–109.

Chapter 12

Bioremediation to Restore the Health of Aquaculture Pond Ecosystem

A. Venkateswara Rao

Introduction

During the past 20 years, aquaculture industry has been growing tremendously, especially that of marine fish, shrimps and bivalves in addition to freshwater fish and prawn. Penaeid shrimps are among the most important and extensively cultured crustaceans in the world (> 60 countries). In 2002, the world's shrimp farmers produced an estimated more than 1.0 million metric tons of whole shrimp and feed mills around the world produced approximately 1.5 million metric tons of shrimp feed.

Shrimp culture all over the world has been frequently affected by viral and bacterial diseases inflicting huge losses and annual losses due to disease has been estimated to be more than US $ 5 billions. Pathogenic microorganisms implicated in these outbreaks were viruses, bacteria, algae, fungi and protozoan parasites. Although diseases pose a serious threat to penaeid shrimp aquaculture, the production of these highly valued crustaceans continue to grow. For preventing and controlling diseases, particularly in aquaculture, the best method is the improving health of culture organisms and elimination of pathogens by improving aquatic environment. Thus, how to improve the ecological environment of aquaculture has become the focus of attention of international aquaculture. Now, researchers have proved the use of probiotic bacteria in aquaculture to improve water quality by balancing bacterial population in water and reducing pathogenic bacterial load. Researchers are increasingly paying more attention to this new approach (ecological aquaculture), and have made considerable headway. In addition to this each year, globally, microbial pathogens cause millions of cases of foodborne disease and result in many hospitalizations and deaths (Jordan Lin *et al.*, 2005).

Competition for survival can be strong, even on a scale invisible to the human eye. Indeed, aquaculture pond bacteria so small that 10,000 fit on a pinhead are subject to the same law of evolution as that Darwin documented on the Galapagos, that is, survival of the fittest or competitive exclusion.

Competitive exclusion is one species out-competing another in a natural search for habitat dominance. As old as time, competitive exclusion is providing science with a neat new means of addressing health and environmental challenges now. The use of competitive exclusion for improving a specific ecology is called "probiotics" (Parkar,1974). Probiotics therapy intentionally introduces strains of beneficial bacteria in order to replace bad bacteria. The first application of probiotics in aquaculture seems relatively recent (Kozasa, 1986), but the interest in such environment-friendly treatments is increasing rapidly.

Aquatic animals are quite different from the land animals for which the probiotic concept was developed, and a preliminary question is the pertinence of probiotic applications to aquaculture. Man and terrestrial livestock undergo embryonic development within an amnion, whereas the larval forms of most fish and shellfish are released in the external environment at an early ontogenetic stage. These larvae are highly exposed to gastro-intestinal microbiota-associated disorders, because they start feeding even through the digestive tract is not yet fully developed (Timmermans, 1987), though the immune system is still incomplete (Vadstein, 1997). Thus, probiotic treatments are particularly desirable during the larval stages.

The transience of aquatic microbes is legitimate the extension of the probiotic concept to living microbial preparations used to treat aquaculture ponds. Moriarty (1997,1998) proposed to extend the definition of probiotics to microbial water additives.

In 1991, Porubcan reported on two attempts at bacterial treatments to improve water quality and production yield of *Penaeus monodon*. (1) Pre-inoculated with nitrifying bacteria decreased the amounts of ammonia and nitrite in the rearing water. This treatment increased shrimp survival (Porubcan, 1991a).(2) The introduction of *Bacillus* spp. in proximity to pond aerators reduced chemical oxygen demand, and increased shrimp harvest (Porubcan,1991b).Moriarty (1998) noted an increase of shrimp/ prawn survival in ponds where some strains of *Bacillus* spp. were introduced. The actual data of Moriarty (1998) showed the inhibitory activity of *Bacillus* spp. against luminous vibrio sp. in pond sediment, but the effect on shrimp/prawn survival might be due either to a probiotic effect, or to an indirect effect on animal health. For instance, the degradation of organic matter by *Bacillus* spp. might improve water quality.

The probiotic treatments may be considered as methods of biological control, the so-called "biocontrol" that termed the limitation or the elimination of pests by the introduction of adverse organisms, like parasites or specific pathogens. Maeda *et al.* (1997) proposed to designate as biocontrol the methods of treatment using "the antagonism among microbes through which pathogens can be killed or reduced in number in the aquaculture environment". Another terminology should designate the applications of nitrifying bacteria that are related to the bioremediation concept. This concept refers to the treatment of pollutants or waste by the use of microorganisms that break down the undesirable substances (Figure 12.1).

Sugita *et al.* (1998) isolated a strain of *Bacillus* sp. that was antagonistic to 63 per cent of the isolates from fish intestine. Pathogenic strains of *Vibrio* or *Aeromonas* have been targeted in most in vitro tests. Treatment with *Lactobacillus brevis* and lactic acid reduced the load of *Vibrio alginolyticus* in the *Artemia* culture water (Villamil, *et al.*, 2003). Some bacteria are antagonistic to viruses (Kamei *et al.*, 1987, 1988; Direkbusarakom *et al.*, 1998), and they may be efficient for the biocontrol of viral diseases (Maeda *et al.*, 1997) and some other probiotics are effective in controlling bacterial diseases in fishes (Aubin, *et al.*, 2005).

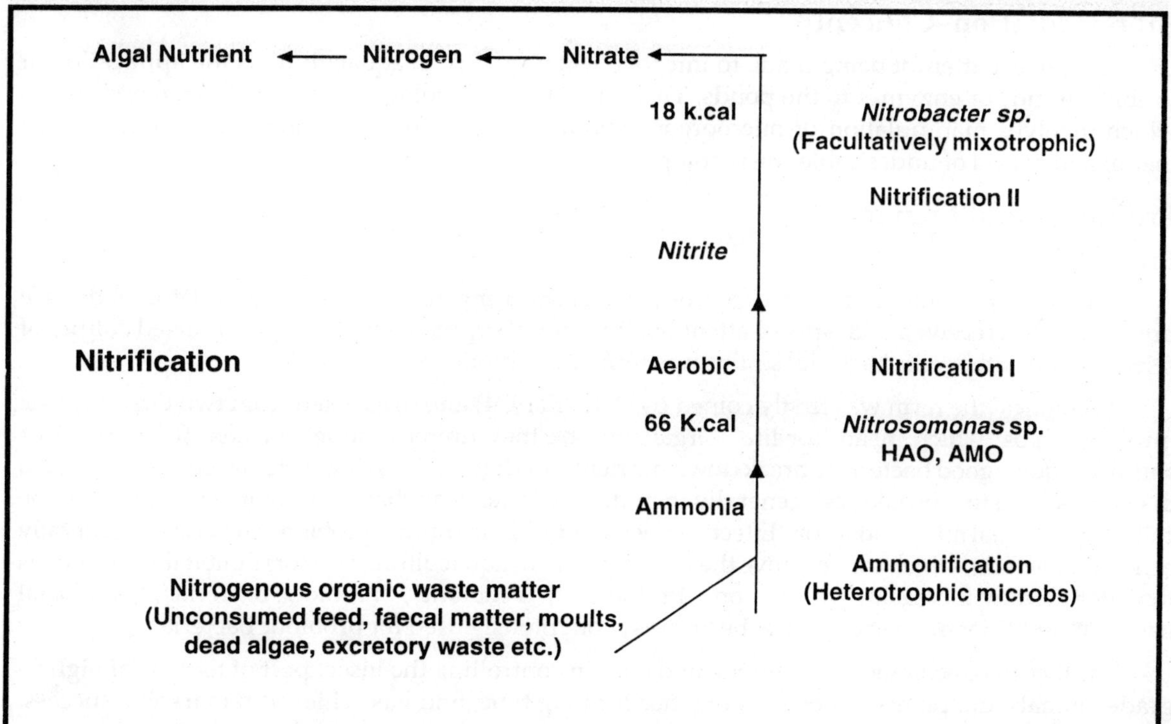

Figure 12.1: Nitrification

The resistance cause by probiotics to infections to shrimp, prawn and fish is due to–Bacterial antagonism, bacterial interference, barrier effect, colonization resistance and competitive exclusion in addition to improving biodegradation of waste organic matter through nitrogen cycle.

To best determine, which bacteria make good probiotics for aquaculture production, scientists scoop up quantities of water from healthy habitat in the ocean where species (aquaculture organisms like shrimp) perform at peak capacity. They analyze the water's makeup and culture single or mixed bacteria they find there to use as probiotics. Their culture does not involve any sort of chemicals or toxins. These are fortified with naturally occurring phytoplankton, amino acids, a wide range of vitamins and minerals, an important variety of anti-oxidants and proteolytic enzymes, as well as an array of nucleic acids. These are the paramount keepers of the code of life, which appears to be in charge of growth and continuous cell repair.

The hypothesis of stimulation of the immune system of aquatic organisms may be also considered. Many immunostimulants have been tested on fish and shellfish, and some of the originated from microbial cell walls, eg. glucans (Anderson 1992,Huang and Song 1999 Huang 2000; Bricknell and Dalmo 2005). It is possible that autochthonous microbiota may stimulate the immune response of aquatic animals to enteric pathogens, as reported in shrimp (Gullian, M., *et al*, 2004) and in land animals (Gaskins, 1997).

Bioremediation–Concept

The newest attempt being made to improve water quality in aquaculture is the application of probiotics and/or enzymes to the ponds. This type of biotechnology is known as *"bioremediation"*, which involves manipulation of microorganisms in ponds to enhance mineralization of organic matter and get rid of undesirable waste compounds.

Probiotics and Enzymes

Probiotics

The concept of biological disease control, particularly using microbiological modulator for disease prevention, has received widespread attention. A bacterial supplement of a single or mixed culture of selected non-pathogenic bacterial strains was termed probiotics.

'Probiotics' the term was firstly coined by Parker (1974) and originated from two Greek words 'pro' and 'bios' which mean 'for life'. Organisms–be they human, cattle, chicken, fish, prawn or shrimp–require good bacteria to break down nutrients for digestion. Living systems require bacteria to decompose waste. "Probiotics" generally includes bacteria, cyanobacteria, fungi, etc. They may be called as "Normal micro biota" or "Effective micro biota". In literature, probiotic bacteria are generally called the bacteria, which can improve the water quality of aquaculture, and (or) inhibit the pathogens in water thereby increasing production. "Probiotics", "Probiont", "Probiotic bacteria", "Beneficial bacteria", or "Friendly bacteria" are the terms synonymously used for probiotic bacteria.

The theory of ecological prevention and cure in controlling the insect pest of terrestrial higher-grade animals and plants has been in practice for long time, and has achieved remarkable success. The bio-controlling theory has been applied to aquaculture and many researchers attempt to use some kind of probiotics in aquaculture water to regulate the micro flora of aquaculture water, control pathogenic microorganisms, to enhance decomposition of the undesirable organic substances in aquaculture water, and improve ecological environment of aquaculture. In addition, the use of probiotics can increase the population of food organisms, improve the nutrition level of aquacultural animals and improve immunity of cultured animals to pathogenic microorganisms.

According to some recent publications, in the aquaculture the mechanism of action of the probiotic bacteria have several aspects.

1. Probiotic bacteria competitively exclude the pathogenic bacteria or produce substances that inhibit the growth of the pathogenic bacteria (*e.g.* Bacitracin and polymyxin produced by *Bacillus* sp.).
2. Provide essential nutrients to enhance the nutrition of the cultured animals.
3. Provide digestive enzymes to enhance the digestion of the cultured animals.
4. Probiotic bacteria directly uptake or decompose the organic matter or toxic material in the water improving the quality of the water.

Hence, the probiotics can decompose the excreta of fish or prawns, remaining food materials, remains of the plankton and other organic materials to CO_2, nitrate and phosphate. These inorganic salts provide the nutrition for the growth of micro algae, while the bacteria grow rapidly and become the dominant group in the water, inhibiting the growth of the pathogenic microorganisms.

The photosynthesis of the micro algae provide dissolved oxygen for oxidation and decomposition of the organic materials and for the respiration of the microbes and cultured animals. This kind of cycle improves the nutrient cycle, and it can create a balance between bacteria and micro algae, and maintaining a good water quality environment for the cultured animals.

Table 12.1: Types of Probiotics

Non-viable probiotics	These are dead.
Freeze-dried probiotics	These will die rapidly upon leaving refrigeration.
Fermentation probiotics	These are produced through fermentation.
Viable probiotics	This is live with guaranteed shelf life, guaranteed number of organisms, have a protocol for counting and to be very stable and efficacious. (produce many benefits).

Table 12.2: Probiotics and Their Role

Bacillus sp.	→ Mineralization and Breakage of proteins
Nitrosomonas sp.	→ Oxidation of ammonia
Nitrobacter sp.	→ Oxidation of nitrites
Aerobacter sp.	→ Reduction of organic matter
Cellulomonas sp.	→ Breakage of plant material

Beneficial Effects of Probiotics may be Mediated by:

1. Neutralization of toxin
2. Suppression of viable count
3. Production of antibacterial compounds
4. Competition for adhesion sites
5. Alternation of microbial metabolism
6. Stimulation of immunity in the host
7. Accelerate the sediment decomposition by producing organic acids
8. Production of hydrogen peroxide
9. Production of enzymes

Application of Probiotics in Aquaculture

1. To regulate the microflora of aquaculture water.
2. To control pathogenic microorganisms.
3. To enhance decomposition of the undesirable organic substances in aquaculture water and improve ecological environment of aquaculture by minimizing the toxic gases like ammonia, nitrite, hydrogen sulfide, methane etc.
4. To increase the population of food organisms.
5. Improves the nutrition level of aquaculture animals and improve immunity of cultured animals to pathogenic microorganisms.
6. The frequent outbreaks of diseases can be prevented.

Enzymes

Enzymes are organic catalysts formed naturally in living cells. They are compounds, which accelerated the rate at which chemical reactions occur and they remain unchanged after the reaction is completed. Thousands of enzymes exist in nature and are responsible for life.

Enzymes work by breaking the chemical bonds that hold compounds together, releasing smaller, more readily absorbed compounds (Julio, H. Cordova-Murueta, *et al.*, 2004). For example, proteases work on proteins, amylases work on starches, cellulases work on cellulose and lipases work on lipids or fats. Recall that cellulose is the major cell wall material in plants and is therefore quite durable.

Nitrogen Cycle

Ammonia is the principal excretory product of most aquatic organisms. Inputs of ammonia cannot be eliminated from the water body. But ammonia is toxic, acutely and chronically, to fish and invertebrates thus it is a critical water quality factor. Ammonia should be maintained below 0.1 mg/L (total ammonia). The most efficient way to do this is by the establishment of a biological filter. A biological filter is a collection of naturally occurring bacteria, which oxidize ammonia to nitrite, and other bacteria, which then convert nitrite to nitrate. Nitrite is formed either by the oxidation of ammonia (nitrification) or the reduction of nitrate (denitrification). Nitrite is toxic to fish and some invertebrates and should be maintained below 0.1 mg/L. It is also a critical water quality factor.

Nitrate is the end product of nitrification. The vast majority of aquaculture ponds accumulate nitrate as they do not contain a denitrifying filter. In general, nitrate should be maintained below 50 mg/L (measured as NO_3-N) but it is not a critical water quality factor.

The most common ways to reduce nitrate are water changes and growing live plants. More sophisticated systems such as denitrifying filters are also available. A denitrifying filter creates an anaerobic region where anaerobic bacteria can grow and reduce nitrate to nitrogen gas. But they can be complex and easily disturbed which kills the anaerobic bacteria. A poorly run denitrifying filter does not convert nitrate all the way to nitrogen gas but instead produces nitrite.

The nitrogen 'cycle' is the oxidation of ammonia to nitrite by bacteria of the genus *Nitrosomonas* and the subsequent oxidation of the nitrite to nitrate by bacteria belonging to the genus *Nitrobacter*. It is easiest to visualize the nitrogen cycle as an endless loop divided into four phases, as follows:

1. Fish, prawn and shrimp excrete ammonia as waste from their gills, kidneys and normal respiration. Ammonia also develops from unconsumed feeds, shell moults of prawn and shrimp, dead algae, zooplankton etc by the microbial activity.
2. A species of bacteria called *Nitrosomonas* converts this ammonia into nitrite
$$(2NH_4^+ + 3O_2 = 2NO_2^- + 4H^+ + 2H_2O)$$
3. A second species of bacteria called *Nitrobacter* converts this nitrite into nitrate
$$(2NO_2^- + O_2 = 2NO_3^-)$$
4. Algae and aquatic plants utilize nitrate to produce chlorophyll, which are in turn consumed by zooplankton and then by fish, prawn and shrimp. Thence the cycle repeats.

These bacteria are important to aqua farmers because without them it is difficult to maintain healthy environmental conditions in the aquaculture ponds.

The biochemical reaction of *Nitrosomonas* sp.: Energy generation in Nitrosomonas by two enzymes, Ammonia monooxygenase (AMO) and Hydroxylamine oxidoreductase (HAO) are involved in the oxidation of ammonia to nitrite:

$$NH_3 + O_2 + 2H^+ \xrightarrow{\text{AMO}} NH_2OH + H_2O$$

$$NH_2OH \xrightarrow{\text{HAO}} NO_2^-$$

The biochemical reaction of *Nitrobacter* is a very simple reaction, involving the cytochrome system as follows:

$$NO_2^- \longrightarrow NO_3^-$$

Nitrobacter sp. is facultatively mixotrophic and capable of growing anaerobically with nitrate as electron acceptor, producing nitrite, nitric oxide and nitrous oxide and then to nitrogen gas. Aerobically it oxidizes nitric oxide to nitrite and thence to nitrate. All tricarboxylic acid cycle enzymes are present (eg. Carboxysomes, poly-beta-hydroxybutyrate (PHB) and polyphosphate granules).

The nitrifying bacteria are gram-negative, non acid-fast rods, which may be pleomorphic or coccoid (Nitrobacter). They may be motile. The main component of the gram-negative cell wall is lipopolysaccharide (LPS). Additionally there is present phospholipid, protein, lipoprotein and a small amount of peptidoglycan. Hence, the component of cell wall of most gram-negative bacteria is associated with endotoxic activity, with which are associated the pyrogenic effects of gram-negative infections like vibriosis. Sixty-six kilocalories of energy are liberated per gram atom of ammonia oxidized. Eighteen kcal of energy is liberated per gram atom of nitrite oxidized.

Nitrification is an obligate aerobic, oxidizing process. This means that it can only occur in an environment, which contains oxygen and the process produces electrons. In general denitrification is a process of bacteria converting nitrate to other substances. It is an anaerobic, reducing process. This means that it occurs in environments without oxygen and the process accepts electrons.

Denitrification is defined as the transformation of nitrate to dinitrogen. Dinitrogen is a gas that is harmless and will bubble out of the system. Between the starting product (nitrate) and the end product (dinitrogen) there are three intermediate products, these are in the order in which they are produced nitrite (NO_2-), Nitric Oxide (NO) and nitrous oxide (N_2O).

Hence, denitrification, like nitrification, is a multi-step process with many intermediate compounds produced before the final product is generated. In most cases these intermediate products are toxic like ammonia and nitrite. Thus denitrification does proceed to complete its process and to expel dinitrogen gas.

It is worth to mention that the nitrate, which is the end product of nitrification process is the major nutrient for the growth of primary producers *i.e.* phytoplankton as well as the microbes. Hence, the micro flora and fauna for their growth will utilize the nitrate and makes limit the process of denitrification.

Another difference between nitrification and denitrification is the type of bacteria, which perform the processes. Nitrification is done by what are called autotrophic bacteria. This term means the bacteria get the carbon they need for cell growth from carbon dioxide. Denitrifying bacteria are heterotrophic bacteria. This means they get their carbon from organic carbon sources such as methane, sucrose or glucose.

In addition these probiotics controls the formation of toxic hydrogen sulfide by eliminating the *Desulfovibrio desulfuricans*, the hydrogen sulfide producing bacteria, through antagonism. The probiotics converts the toxic sulfide to nutritive sulphate.

Conclusion

The probiotics can decompose the excreta of shrimp, prawn or fish, remaining food materials, remains of the plankton and other organic materials to CO_2, nitrate and phosphate. These inorganic salts provide the nutrition for the growth of micro algae, while the bacteria grow rapidly and become the dominant group in the water, inhibiting the growth of the pathogenic microorganisms. The photosynthesis of the micro algae provide dissolved oxygen for oxidation and decomposition of the organic materials and for the respiration of the microbes and cultured animals. This kind of cycle may improve the nutrient cycle, and it can create a balance between bacteria and micro algae, and maintaining a good water quality environment for the cultured animals.

Probiotics reduce the level of infection and mortality in aquaculture organisms. Probiotics compete with the disease-causing bacteria for food. Customarily, ponds have indigenous bad bacteria that leisurely eat on an abundance of undigested food, faecal matter, and dead algae. When farmers suddenly introduce a large quantity of these good bacteria–probiotics–into a pond, the newcomers eliminate the pre-existing bad bacteria out of the nutrient queue. The old bacteria, often-bad bacteria, never having had to compete for food, cannot keep pace with the aggressive probiotics.

Probiotics excretions make the pond medium less inhabitable for bad bacteria. Not only do probiotics bacteria have terrific appetites; they excrete enzymes–exoenzymes–as a natural byproduct of their metabolic activity, just as human's sweat. The enzyme excretions infuse and spread throughout the pond medium, changing its chemistry, resulting destroy of bad bacteria. Hence, probiotics therapy greatly improves pond water quality and reduces the pollution level of effluent before its release into the environment.

Probiotics speed the breakdown of organic waste fragments (dissolved proteins and unused feed), thus lessening sludge build-up. If sludge is not removed or does not decompose, dangerous concentrations of sulfide, nitrite, ammonia and various organic acids can occur. While the exo-enzymes of good bacteria go to work on larger waste particles and breaking them down through chemical reaction. As a result, there are fewer harmful chemicals in the medium.

Probiotics balance algae growth through providing nutrients by their bio-degradation activity. Probiotics utilizes phosphate for their body metabolic activities and hence prevents over growth (heavy blooms) of algae by limiting the phosphate. Dead algae are one of bacteria's favorite foods. Also probiotics diminish nutrients normally consumed by algae and results less slime in the ponds.

Probiotics make better habitat for culture organisms. With less accumulation of organic matter on the pond bottom, more oxygen can penetrate the sediment. For eg. Prawn and shrimp characteristically burrow in the sediment. By loosening the sediment, probiotics make this burrowing easier. Moreover, they diminish the toxin level in the sediment itself. Probiotics maximize fish, prawn and shrimp nutrition. In addition to consuming the feed administered by farmers, prawn and shrimp graze on tiny zooplankton that is present in the water supply. Zooplankton eats bacteria. Probiotics are another nutrient for existing zooplankton in the pond medium, thus invigorating the zooplankton population, and strengthen up the food supply to culture organisms.

Probiotics optimize the immune systems of culture organisms, increasing their resistance to disease. By upgrading the pond environment, augmenting the food supply, and making habitat more comfortable, probiotics fortify fish, prawn and shrimp immune system.

The application of probiotics in aquaculture shown promise, but further investigations may be expected with propagation of molecular approaches to analyse bacterial communities (Raskin *et al.*, 1997; Wallner *et al.*, 1997; Hugenholtz *et al.*, 1998; Gatesoupe, 1999).

References

Anderson, D.P., 1992. Immunostimulants, adjuvants and vaccine carriers in fish: applications to aquaculture. *Annu. Rev. Fish Dis.*, 2: 281–307.

Aubin, J. *et al.*, 2005. Trial of probiotics to prevent the vertebral column compression syndrome in rainbow trout (*Oncorhynchus mykiss* Walbaum). *Aquaculture Research*, 36: 758.

Bricknell, I. and Dalmo, R.A., 2005. The use of immunostimulants in fish larval aquaculture. *Fish and Shellfish Immunology*, 19: 457–472.

Direkbusarakom, S., Yoshimizu, M., Ezura, Y., Ruangpan, L. and Danayadol, Y., 1998. *Vibrio* spp., the dominant flora in shrimp hatchery against some fish pathogenic viruses. *J. Mar. Biotechnol.*, 6: 266–267.

Gaskins, H.R., 1997. Immunological aspects of host/microbiota interactions at the intestinal epithelium. In: *Gastrointestinal Microbiology, Vol. 2, Gastrointestinal Microbes and Host Interactions*, (Eds.) Mackie, R.I., With, B.A., Isaacson, R.E. Chapman and Hall Microbiology Series, International Thomson Publishing, New York, pp. 537–587.

Gatesoupe, F.J., 1999. The use of probiotics in aquaculture. *Aquaculture*, 180: 147–165.

Gullian, M., Thompson, F. and Rodriguez, J., 2004. Selection of probiotic bacteria and study of their immunostimulatory effect in Penaeus vannamei. *Aquaculture*, 233: 1–14.

Huang, C.C. and Song, Y.L., 1999. Maternal transmission of immunity to white spot syndrome associated virus (WSSV) in shrimp (*Penaeus monodon*). *Developmental and Comparative Immunology*, 23: 545–552.

Hugenholtz, P., Pitulle, C., Hershberger, K.L. and Pace, N.R., 1998. Novel division level bacterial diversity in a Yellowstone hot spring. *J. Bacteriol.*, 180: 366–376.

Jordan Lin, C.T. *et al.*, 2005. Awareness of foodborne pathogens among US consumers. *Food Quality and Preference*, 16: 401–412.

Julio, H. Cordova-Murueta *et al.*, 2004. Effect of stressors on shrimp digestive enzymes from assays of feces: an alternate method of evaluation. *Aquaculture*, 233: 439–449.

Kamei, Y., Yoshimizu, M., Ezura, Y. and Kimura, T., 1987. Screening of bacteria with antiviral activity against infectious hematopoietic necrosis virus (HNV) from estuarine and marine environments. *Bull. Jpn. Soc. Sci. Fish*, 53: 2179–2185.

Kamei, Y., Yoshimizu, M., Ezura, Y. and Kimura, T., 1988. Screening of bacteria with antiviral activity from fresh water salmonid hatcheries. *Microbiol. Immunol.*, 32: 67–73.

Kozasa, M., 1986. Toyocerin (*Bacillus toyoi*) as growth promoter for animal feeding. *Microbiol. Aliment. Nutr.*, 4: 121–135.

Maeda, M., Nogami, K., Kanematsu, M. and Hirayama, K., 1997. The concept of biological control methods in aquaculture. *Hydrobiologia*, 358: 285–290.

Moriarty, D.J.W., 1997. The role of microorganisms in aquaculture ponds. *Aquaculture*, 151: 333–349.

Moriarty, D.J.W., 1998. Control of luminous *Vibrio* species in penaeid aquaculture ponds. *Aquaculture*, 164: 351–358.

Parker, R.B., 1974. Probiotics. The other half of the antibiotics story. *Anim. Nutr. Health*, 29: 4–8.

Porubcan, R.S., 1991a. Reduction of ammonia nitrogen and nitrite in tanks of *Penaeus monodon* using floating biofilters containing processed diatomaceous earth media pre-inoculated with nitrifying bacteria. *Program and Abstracts of the 22nd Annual Conference and Exposition*, 16–20 June, San Juan, Puerto Rico. World Aquaculture Society.

Porubcan, R.S., 1991b. Reduction in chemical oxygen demand and improvement in *Penaeus monodon* yield in ponds inoculated with aerobic *Bacillus* bacteria. *Program and Abstracts of the 22nd Annual Conference and Exposition*, 16–20 June, San Juan, Puerto Rico. World Aquaculture Society.

Raskin, L., Capman, W.C., Sharp, R., Poulsen, L.K. and Stahi, D.A., 1997. Molecular ecology of gastrointestinal ecosystems. In: *Gastrointestinal Microbiology, Vol. 2, Gastrointestinal Microbes and Host Interactions*, (Eds.) Mackie, R.I., With, B.A. and R.E. Isaacson. Chapman and Hall Microbiology Series, International Thomson Publishing, New York, pp. 243–298.

Song, Y.L. and Huang, C.C., 2000. Application of immunostimulants to prevent shrimp diseases. In: *Recent Advances in Marine biotechnology, Vol.5, Immunobiology and Pathology*, p. 173–187.

Sugita, H., Hirose, Y., Matsuo, N. and Deguchi, Y., 1998. Production of the antimicrobial substance by *Bacillus* sp. strain NM 12, an intestinal bacterium of Japanese coastal fish. *Aquaculture*, 165: 269–280.

Timmermans, L.P.M., 1987. Early development and differentiation in fish. *Sarsia*, 72: 331–339.

Vadstein, O., 1997. The use of immunostimulation in marine larviculture: possibilities and challenges. *Aquaculture*, 155: 401–417.

Villamil, L., Figueras, A., Planas, M. and Novoa, B., 2003. Control of *Vibrio alginolyticus* in *Artemia* culture by treatment with bacterial probiotics. *Aquaculture*, 219: 43–56.

Wallner, G., Fuchs, B., Spring, S., Beisker, W. and Amman, R., 1997. Flow sorting of microorganisms for molecular analysis. *Appl. Environ. Microbiol.*, 63: 4223–4231.

Chapter 13

Development of Fisheries through Biotechnology: An Appeal

Indranil Ghosh

Introduction

The State of West Bengal is regarded as the highest producer of food fish in India, with per capita fish consumption of 15.6Kg against the national average of 8Kg and the world average of 13Kg. It is bestowed with remarkable variation of various Physiographic characters and varied environment with huge water resources suitable for commercial scale production of variety of fishes. Despite all, optimum production is far away from at hand due to some lack of adoption of proper technology in some field. The present article focuses the scope of further development in this sector through the application of some advanced method of biotechnology easily adoptable to the fisheries personnel in the country that are currently being taken up.

Eco-Geographical Conditions for Fisheries

West Bengal is probably the only state having six major agro-climatic regions amongst thirteen zones in India. The state is situated with Lat. 21°38¢N to 27°10¢ N and Long 85°38¢E with a geographical coverage area of 87,853 sq km. The state is endowed with remarkable variations of physiographic resources from the sea to snow having elevation 5m. in the South to 3658m. from main sea level in the North. The state exhibits 3 broad well-marked geoclimatic region and stages of developments: -

1. Himalayan mountain track in the North.
2. The Alluvial and deltaic plain in the Middle and South.
3. The Undulating rocky plateau in the West.

The climate of this state ranges from sub humid and humid to poor humid.There is no denying the fact that the important role played by fisheries and aquaculture in the agrarian economy of West Bengal decisively determines the progress and prosperity in the state, as the state not only possesses vast area of diverse geographical characteristics as stated above, but also the state is bestowed with beautiful natural resources in the form of river systems, estuaries, ponds, tanks, reservoirs and wetlands. It is the state where the World's largest River Delta "The Sunderbans" situated. These water bodies are mostly exposed to warm tropical climate, extremely productive and they harbor an exclusives spectrum of fish genetic resources. Optimum and sustainable utilization of these resources can lead to many fold increase in fish production.

Present Status of Fisheries in West Bengal

The state of W.B. has always attracted attention as the national level for being the highest producer of freshwater table fish and fish seed in the country together with the unique distinction of having the maximum water area under traditional shrimp farming. Apart from that, the state also offers the " Largest Potential Brackish water Resources " for aquaculture (4,05,000 ha) followed by Gujrat (3,76,000 ha), Andra Pradesh (1,50,000 ha), Maharashtra (80,000 ha), and Kerala (65,000 ha). It also engages apporx. 1.1 million fishermen constituting 0.67 per cent of the total population of the state. Moreover, the state incidentally is the highest consumer of fish and fish products with a per capita fish consumption being 15.6 kg. as against the world avg. of 13 kg. and the national avg. of 8 kg. The demand and supply gap is projected to be 1.63 lakh tonnes by the Year 2002.The state also provided the major inland fish market in the country and is strategically located for the trade route of inland fish movement from the Southern, Northern and Western markets to the Eastern and North-Eastern markets of the Country.

Freshwater Fisheries Sector

The state offers a potential Freshwater Aquaculture resources of 2.76 lakh ha of which 2.08 lakh ha is already developed leaving a balance water area 68,000 ha yet to be developed. The amount of derelict and semi-derelict area in 0.27 lakh ha and 0.55 lakh ha respectively.

Brackishwater Fisheries Sector

Having the largest resource in the country (4.05 lakh ha),out of the estimated potential aquaculture area being 2.10 lakh ha, only 0.55 lakh has been developed of which 0.34 lakh ha is under traditional systems of 'Bhery farming', most of which are located in North and South 24 Pgs whereas improved farming practices are prevalent in Medinipur district. The state has by and large resources to the extent of 75 per cent (1.55 lakh ha) for future development.

Marine Fisheries Sector

West Bengal could offer a Maximum Sustainable Yield (MSY) of 1.81 lakh tonnes of fish as against the present landing of 1.71 lakh tonnes leaving a balance potential of 9500 tonnes. About 20 per cent of total production of the state comes from this sector. Its inshore traditional Zone in particular has already been optimally exploited, while offshore and deep-sea areas virtually remain unexploited and unexplored.

Some New Dimensions for the Development and Commercialization of Fisheries Sector in the State

1. Vertical expansion of fresh water aquaculture in the lines of other states like Andhra Pradesh and Punjab considering the limitations of horizontal expansion depending upon different agro-climatic conditions.

2. Diversification of freshwater aquaculture adopting freshwater prawn culture including the Giant Freshwater Prawn (*Macrobrachium resenbergii*) as well as other spp.
3. Introductional species like 'Barbs'(*Puntius ticto, P, Sophore*), Mola (*Amblypharyngodon mola*) etc. These fishes are on the verge of extinction as they are having a good demand in the market with obvious nutritive value.
4. Reclamation of marginal areas of beels for developing aquaculture estates and introduction of cage culture in open water.
5. Establishing brackishwater aquaculture estates to provided common infrastructural facilities to provide common infrastructural facilities to small farmers since the increased cost of infrastructure following the Aquaculture Authority Guidelines may not be financially rewarding in case of small holdings.n of high value air breathing species like *Pangassius*, 'Magur' (*Clarias batrachus*) "Singhi" (*Hetropneustes fossils*), Pabda (*Ompak Pabda*), Indian climbing Perch (Locally called 'koi', *Anabas testudeneus*),Freshwater Shark(locally called'Boal', *Wallago attu*) etc.
6. Introducing of other high.
7. Introduction of some newer species like Sea- Bass (*Lates calcarifer*), Milk-Fish (*Chanos chanos*), Grey–Mullets (*Liza parsia, L. tade*) in brackishwater aquaculture sector.
8. Introduction of non-conventional fisheries items like 'Mussels" (*Perna spp.*), Clam(*Paphia sp.*) with shrimp farming.
9. Introduction of proper technology for 'Sea-weed' farming.
10. Introduction of " Mud Carbs" (*Scylla serrata* and *S. tranquebariaca*) will be an alternative opportunity particularly for the districts of North and South 24 Pgs and Medinipur.
11. Seed Production and culture of Ornamental Fishes, both the exotic as well as indigenous species, for their demand in domestic and international markets.
12. Establishment of 'Fish-Feed' manufacturing units both for FW and BW sectors.
13. Introduction of Freshwater Pearl Culture.
14. Aquatic bio-fertilization and organic farming.
15. Introduction of " Cage " and " Pen " culture of Carps.
16. Commercialization and popularization of Integrated Aquafarming (like Duck-cum-Fish culture, Fishery with Piggery, Poultry etc)
17. Establishment of Fish Disease Diagnosis centres and their control.

The application of Biotechnology has given more encouraging results in Agriculture and Animal Husbandry, while its application is very limited in Aquaculture even today. Biotechnology is comparatively a recent innovation as applied aquaculture for enhancing fish production in the country, while its application in fisheries sectors have been proved to be highly beneficial all over the world.

Several Approaches of Biotechnology may lead to the Manifold Increase in Fresh Production in the State of West Bengal

1. Induced breeding techniques for the Fish Seed Production: Using natural and synthetic spawning hormone injection to the brood- fishes.
2. Selective breeding and hybridization of cultured species:

(*a*) To produce species, superior to their parents, with faster growth and enhanced production.

(*b*) To maintain the purity of 'germplasm' and its characters.

3. Polyculture of Compatible Species with Carps: For increased production by proper utilization of all the 'Ecological niche' of water area beyond the carrying capacity of that water body.
4. Running water Culture: To get high production because of intensive stocking.
5. Manipulation of Chromosome Sets: The phenomena like polyploidy, gynogenesis and androgenesis may be introduced to increase production.
6. Cryopresrvation Technology: Programmes may be framed for preservation of gametes for future and to produce fish seeds whenever necessary irrespective of their breeding season using liquid nitrogen.
7. Hormonal Manipulation for Sex-Control: Efforts may be given for production of 'mono-sex' fish by using hormones to increase production. This technique can be widely used in " Ornamental Fishes" also, raising the high price in the market.
8. Ready to eat fish product and value added product development:By utilizing the low value fishes.

Future Projection and Human Resources Developments

In the bright perspective as stated above, a 'Biotechnology park' may be established and equipped with some of the above mentioned items whichever necessary in area specific manner depending upon the agro-climatic conditions, Physico-chemical characteristics of water and soil, domestic demand. It ia is a matter of fact that a number of such biotech parks have already been proposed. Some of the Institutions with various disciplines are coming to take part in those parks. At present though it is in the preliminary stage, but surely promising a lot in the future.

Human Resources Development programmes as different levels of development of fisheries personnel, entrepreneurs, small and marginal farmers are vital to bring about the desired qualitative and quantitative improvement of aquaculture and allied branches. For this purpose, some ' Technical Education Society ' may be established rather than the conventional University/Institution etc., where 'Hands-on-training' through various short-term certificate courses on different arena/field should be offered and skilled professionals may be developed.

The duration of courses may vary from 5 days to a few weeks, depending upon depth and dimension of the particular trade. The number of participants in each group to the tune of 10-12, may be regarded as ideal.

Refresher courses on regular basis may be arranged to train the people with newly invented technologies. The role of electronic media and devices can't be denied.

The society should be equipped with well-established laboratories in each discipline both for the training purpose as well as for the commercial purpose, by analyzing water and soil, identifying the diseases etc. against a reasonable fee within very short time of the emergence of the scourge.

Marketing Channels and Infrastructures Development

1. To tie up with foreign agencies through Govt. organizations for earning foreign exchange.
2. To establish proper market place of particulars products.

3. Lack of proper database has often been great constraint of management measures should be taken in the respect for compiling data on water resources, fish productions levels and other technical details, practicing aquaculturists and population dependent upon aquaculture and other allied branches, marketing channels, frequently asked questions (FAQs) etc. This would greatly facilitate the establishment of a proper information system.
4. Provision of a package of Technical, Financial and Extension support to the farming community/entrepreneurs.
5. Awareness programmers for the people.
6. In order to ensure remunerative returns, it is necessary to organise marketing channels supported by adequate storage facilities that would serve to regulate the market and assure reasonable prices.
7. Budgetary support for various infrastuctural development, training and extension need to be enhanced.
8. Creation of necessary infrastructure facilities for the production of quality fish seed and marketing.
9. Availability of required credit amount from the financial institutions should be assured.
10. Availability of regular cargo space at affordable tariff by rail and air are a must particularly to ensure continuity of the Ornamental Fisheries Trade. Live cargo-holding facility near Airport and Terminal railway Stations involving Oxygen packing, Temporary holding facilities, forwarding and clearing and transportation facilities should be availed.
11. Ornamental Fish show should be organised at least once in a Year.

Chapter 14

Optimization of Resources for Food and Income Security

Salim Sultan

> "The availability of food is a critical factor in the well being of any nation, one that even developed nations cannot afford to ignore. For a country like India with a large population the task of maintaining a regular supply of food in stable, consistent and a viable manner to its people irrespective of vagaries of the weather and other natural calamities, is itself a stupendous task. Measures aimed at ensuring such unhindered supply fall under steps aimed towards food security".
>
> *Dr. A. P.J. Abdul Kalam*

Introduction

The present sweeping socio-economic developments have resulted in many problems that are endangering not only the present and future generations but also equality and equity in development. It is due to maladjustment between human society and its environments which is the greatest problem that we will be facing in near future. This maladjustment is due to natural and anthropogenic factors such as land degradation, draught or famine and their whole nexus of related problems. Green revolution led to the achievement of near self sufficiency in food production. Now it is facing new set of challenges posed by unbalanced production of food grains, excessive use of chemical fertilizers and pesticides. We are faced with an imperative to relate our developments to the external factors of globalization, liberalization and privatization of production and trade regimes. Our priorities today are the need to bring about necessary changes to negate these problems, to enhance production and productivity and to increase the opportunities for off farm employment and value addition through agro-processing and small-medium enterprises. The major challenge before us is how to maintain and enhance the integrity of natural resource base, while at the same time increasing productivity and value addition. Prof. M. Swamminathan, the architect of Green revolution has advocated the "Bio-village" model of rural development with its thrust on proper, pro-job pro-women approach. He calls it evergreen revolution which will bring about radical transformation in the countryside. This concept can inter-realize in our country only through a massive land care/use movement for bringing about rural assistance.

Nutritional Scenario

Most people associate agriculture with food grains only. But food grains account to only 26 per cent of agriculture GDP, followed by fruits/vegetables (21.3 per cent) and dairying (17.7 per cent). Fisheries account for 5.5 per cent and their share is rising. Between 1991 and 2003 food grain production rose only 30 per cent while aquaculture output rose 102 per cent.One third population of India is still below poverty line consuming only protein deficient starchy diet and cannot meet the basic nutritional standards. This wide spread poverty induced protein deficient diet from malnutrition has created need for enhancing food production to cope with ever increasing human population. This situation has assumed vital importance all over the world especially in more populated and resource poor countries. In this context aquaculture *i.e.* use of aquatic resources for development purposes at par with terrestrial resources for agriculture and animal husbandry is required. Fish is a valuable and easily accessible source of food, rich in protein. Like rice, wheat and pulses, fish is also a staple diet for about 30 per cent population particularly economically weaker section who can not afford high value nutritionally rich food items. Low grade fish generally serve their purpose. The present fish availability is only 25 per cent of prescribed norms to 50 per cent non vegetarian population of our country. With population growing at fast pace and increasing awareness among people on nutritional quality of fish, the production of fish has to be further intensified. "Green Revolution" has facilitated food production through High Yielding Varieties of cereals but most amazingly it has encouraged only starchy food with consequent depletion of proteinaceous pulses. The ratio of sown area for rice + wheat: pulses are 78:22 and for production it is 93.1: 6.9 at all India level (20001). Uttar Pradesh has total land mass of about 242.02 lac hectares, of which 168.19 lac hectares are under cultivation (2000-01). The gross cropped area is 203.98 lac hectares. Pulses occupy only 12 per cent of cropped area (26.83 lac ha) as compared to 81 per cent for wheat and rice. Thus huge protein demand cannot be fulfilled by the land based agro or livestock protein. As incomes rise, grain is increasingly used as feed grain for meat production. Producing one kilo of beef takes seven kilos of grain, but one kilo of fish needs only 1.6–2.0 kilos of grains. So fish is economically and ecologically efficient converters. Much fish in India are not fed at all. Natural organic wastes run in to ponds and help the growth of plankton, which fish eat. Although aquaculture alone can not fill the gape but it becomes more relevant and important in present scenario.

The vision that the cause of food insecurity and vulnerability should be viewed under the over all frame work of livelihood security is gaining wider acceptance of development partners and stake holders. United Nations Millennium Development Goals (MDG) has pledged to halve the proportion of worlds poor and of people who suffer from hunger by 2015. Given the circumstances of our country, India has a special responsibility in this regard (Singh, 2005). Persistent malnutrition in our society could potentially end up with people of weakened bodies with impaired brain development, psycho-emotional disturbances, antisocial responses, compromised immune systems leading to degenerative diseases, coronary heart diseases and consequent irrational instincts and turmoil of life. Hence development and promotion of fisheries is as needed as other farm sectors to serve the needy masses.

Fisheries

Fisheries sector occupies an important place in the socio economic development of the country. It is a source of livelihood to over 14 million people and earner of foreign exchange to the tune of Rs. 6000 crores per annum. India is the fourth largest producer of fish in world preceded by China, Peru and Japan and second largest producer of fresh water fishes preceded by China as per FAO statistics 2000. Fisheries was once dominated by ocean fisher flocks, but over fishing has led to virtual stagnation of the sea catch in the past decade. A meanwhile inland fishery has grown fast and now account for 55 per cent of the total catch (7 million tones) yet performance lags behind potential. India's global share is about 10 per cent against China's 57 per cent. India's plentiful bounty of nature and biodiversity in regards to fish germplasm resources are amazing. While about 25426 fish species are believed to inhabit water bodies all over the world, we know 2500 species (fresh water 930 + marine1570) in India. About 20 per cent of Indian species are considered to be economically important.

Figure 14.1: Fish Harvest from Pond

Centuries of experience have taught the Chinese to maximize the use of pond water. In each pond they grow four carp varieties which feed at different levels. Silver and big carps are top feeders, grass carps are middle feeders and common carp are bottom feeders; such multi level aquaculture is attempted in India too through state fishery departments. Culture technology established in India still gives an average production of 2100 kg/ha/yr. This present average production can be stepped up several folds moving towards global productivity levels. Intensive fish production technology by CIFA (10-15 t/ha/yr) is commendable but still to be transferred for adoption by resource poor farmers. Extension and seed supply efforts of state fisheries departments are woefully inadequate due genuine multiple constraints. Despite of all this fish production is rising but it faces limits. Thanks to free electricity farmers are pumping water like mad and so water in all ponds and tanks are falling sharply. Few tanks and ponds have perennial water which is needed for fisheries. Farmers can pump water to fill their fish tanks but those using diesel pumps say it is prohibitively expensive.

Table 14.1: Comparative Fishery Resource Potential

Category	*India*	*UP*
EEZ	2.02 m km^2	-
Brackish water (lac ha)	12.40	-
Rivers and canals (km)	195210	28500
Reservoirs (lac ha)	29.07	1.38
Lakes (lac ha)	7.98	1.33
Ponds (lac ha)	24.14	1.61
Marine production "000 t	2900	-
Inland production "000 t	3457	267
Growth rate inland sector		

Land Resource

To achieve the projected scenario, we would have to judiciously utilize various resources in implementing programs with a singular aim of raising proteinaceous production through land use, creation of infrastructure, upgraded utilization levels, propagation of extensive as well as intensive aquaculture in ponds and tanks, development management of reservoirs, conservation and management of capture fisheries from rivers and streams and innovated controlled culture of fish in various other aquatic–ecosystems such as canals, lakes, derelict waters and dead river courses. Land use prescriptions are mainly determined by the physical attributes, specially climate, topography and soil, human attributes of an area which include social structures, cropping patterns, way of life and even the traditions, culture habits and temperament of inhabitants play a crucial role in the determination of land use. The term land use not to be understood merely in sense of allocation process *i.e.* allocation of land for different uses but must be seen in a wider context of inclusive development whose bottom line is defined by economics, ecology and equity. The land use debate is going on with different properties and aims. The soil and space scientists look at it as a prescriptive problem in the context of climate, soil and topography. Technologists on the other hand, look at it as a problem of recreating nature through biotechnology and genetic engineering. Ecological scientists are concerned about reducing damage to our ecosystems. Resource management specialists are concerned with conservation problems and the allocation of real costs of resource exploitation. Spatial analysts are concerned about achieving more rational patterns of economic activities. Behavioral scientists seek to reduce the unnecessary constraints on people and to improve their ability to make effective decisions. Economists would like the problem to be considered with in the global economic system and tend to view it in a historical perspective linked to economical and political structures. Since it is the man who is putting the land and its biotic resources to their different uses it has to be seen in the context of interface of the earth and man. This contextual space refers not only to the visible elements of geographic space *i.e.* Physical features, land reforms and natural land resource elements but also relatively non visible human values such as social relations, cultural hopes and aspiration of local community. Land use must be seen in the context of improving our rural livelihoods in accordance with human capabilities and cultural traits. Land care is basic to harness our bio-diversity and to realize the potential of our production system on a sustained basis. In our present context of mounting demographic pressure and challenges, related to sustainable enhanced agriculture production, it calls for vital changes and departures in our mind set including values, economics and community endeavors different from those that prevailing today. Our policy should seek to address issues such as role of bio-technology, agricultural technologies in different ecosystems, biodiversity and institutional reforms.

Land Care

In 1960's Karnataka initiated a land care movement by the name of land army. Today with a better organizational set-up, it is perhaps worth giving a push to this idea of land army for land development. Although formulated in 1990's, state of UP has taken up this task on priority basis now by formulating action plan for three years (2005-08). Important activities to be covered under this plan are 1) Improvement of degraded and fallow land 2) Productivity enhancement in these treated areas by encouraging agriculture, horticulture and agro forestry 3) improvement of ponds for aquaculture 4) Improvement of waterlogged areas to convert these in to productive mode. It is a participatory project for farmers and will cover 100 thousand hectare land including 1000 ha ponds in three years with a total cost of about Rs.100 crores.

Honorable Apex Court's Directive

In a land mark decision (Hinch Lal Tiwari vs. Kamla Devi and others, civil appeal number 4787 decided on July 25, 2001), honorable Supreme Court issued following directives. "It is important to notice that the material resources of the community like forests, tanks, ponds, hillocks, mountain etc are nature's bounty. They maintain delicate ecological balance. They need to be protected for a proper and healthy environment which enables people to enjoy a quality life which is the essence of the guaranteed right under article 21 of the constitution. The government including the revenue authorities *i.e.* Respondent 11-13 having noticed that a pond is falling in disuse, should have bestowed their attention to develop the same which, would on one end, have prevented ecological disaster and on the other provided better environment for the benefit of the public at large. Such vigil is the best protection against the knavish attempts to seek allotments in non abadi sites". Honorable court further observed that "The state including respondent 11 to 13 shall restore the pond, develop and maintain the same as a recreational spot which will undoubtedly be in the best interest of villagers. Further it will help in maintaining ecological balance and protecting the environment in regard to which this court has repeatedly expressed its concern. Such measures must begin at the grass root level, if they were to become the nation's pride (Para 13 and 14). These directives would have far reaching implications in restoring and preservation of water bodies specially with growing land prices.

Water: The Most Precious Natural Resource

Water is the most precious natural resource in the world. Since early days of civilization, all cultures whether human, agriculture or animals have flourished near water. Water is critical for fish; it has non consumptive use in fishery sector but acts as substrate for its growth and sustenance. Now it is a commodity not a medium, therefore every drop of water should be preserved as its availability is falling every day. Drinking, irrigation and power generation get priority over ecosystem preservation in usage of water. The emerging fresh water scarcity is recognized as an important issue. Balancing the needs of aquatic environments and other uses is becoming critical in many rivers systems as population and associated water demands increase. Striking a balance between allocation of water for direct human use (domestic, agriculture, power generation and industry) and indirect human use (maintenance of ecosystem, goods and services) is necessary for productivity enhancement and sustainability (Mishra and Mohanty, 2004). India is privileged country, particularly the sub Himalayan regions, in terms of monsoon rains as it receives huge quantity of water annually which helps in recharging groundwater table also. Storage, conservation and subsequent use of water are very critical to link-all related activities in a composite manner to ensure optimum utilization of this resource. Therefore, any water management issue does affect fishery directly.

Water Wealth in Uttar Pradesh

UP is endowed with three types of water resource, rain water, surface water and underground water. Another hidden portion is humidity which considerably affects biotic ecosystems. State receives about 23.54 m ha meter rain water out of that about 37.50 per cent flows down through rivers. Surface water flow in Ganga basin is about 55.03 m ha meter out of which state share is 33.55 m ha meter while usable portion is only 19.95 m ha meter. Water storage in the under ground aquifers of the state is 9.51 m ha meter (Anon, 2003). These figures reflect enormous water which can be brought under production cycle through appropriate management after initial investment.

Rivers

Fish catch from rivers is declining due to over fishing in some cases while environmental stress in others. Natural hydrographs in rivers are altered by the construction of dams and diversion of water for agriculture, industry and community use. Over the last fifty years almost 250 billion cubic meters of storage has been created intercepting 30 per cent of the available surface flow. Almost 70-80 per cent of annual flow is diverted in to few projects. These trends are unlikely to be reversed. And this would imply that fishery would be placed under increasing pressure. The corrective measures can be effective only when holistic approach to fisheries management is adopted, whereby the land and water resources necessary to fishery are negotiated with other users of rivers (Vass, 2005).

Reservoirs

Reservoirs are men made impoundments meant for irrigation, power generation and domestic use. Both fluviatile and lacustrine system exists side by side. In the present day fishery sector this area has opened new vistas with enormous fish production potential. Indian reservoirs with a total water spread area of over 3 million hectares are considered as an invaluable resource for enhancing inland fish production. The current level of fish yield is low, averaging about 20 kg $ha^{-1}yr^{-1}$. These under utilized fishery resources offer immense scope and potential for generating additional income by adopting simple management measures. On an average the production from large reservoirs can be raised to 50-100 kg $ha^{-1}yr^{-1}$ and from small reservoirs to 100-300 kg $ha^{-1}yr^{-1}$amounting to total fish 2.45 lac tones annually (Krishna Rao, 2005.)

Canals

To increase the fish acreage conversion of all canals in to fisheries seems to be a pliable proposition. Every canal can be divided in to 100 meter sections separated by aluminum mesh (which will not rust). And each section can be leased out to local farmers. Water for irrigation will not be disturbed by this arrangement, and will actually be encircled by fish faeces, which are fertilizer. Irrigation department will earn extra revenue which they badly need. Aqua culturist will treat their canals as ponds, choosing what fish to grow and when to harvest their catch. India has hundreds of kilometers of canals which can produce millions of tones of fish by simply bringing existing canal area under culture (Aiyar 2005).

Lakes/Natural Wetlands

These are flood plain wetlands having their permanent or temporary association with parent river course. They are under severe threat of siltation with increased macrophytic load pushing them further towards non functional stages in a water body resulting in their minimal utilization for production purposes. Amazingly, these are harbinger of aquatic biodiversity and provide a very productive land area for agriculture in catchments. Cleaning weeds, regulation of mesh size, stock enhancement, phase wise siltation and construction of water regulatory structures are some of the intervention options for better utilization of these water bodies. This is an enormous and capital intensive task, therefore, creation of separate wetland board will be a welcome step to formulate strategies for the development and utilization of this resource at desired pace (Sultan, 2004).

Figure 14.2: A Depleted Lake

Pond Fishery

Occurrence of ponds is a very common feature in rural areas. They are part and parcel of daily life with multiple utility in rural culture. Uttar Pradesh possess about 1.61 lac ha land area as ponds out of which about 70 per cent is presently covered under fisheries. Effective utilization through scientific management after desired modification of these aquatic bodies has tremendous scope for employment and fish production. Interestingly, these smaller water bodies provide very productive assets due to increased possibilities of controlled interventions. They function on *put and take out* pattern of production cycles. Therefore, as a consequence of better productivity and huge area coverage, pond fishery contributes a major share in inland fish production among all inland aquatic resources.

Fish Farmer's Development Agencies

This is a centrally sponsored scheme for the renovation and excavation of ponds to make them suitable for fishery. Institutional support from various landing agencies are available to farmers on NABARD approved unit costs. These autonomous bodies have done tremendous job in popularization of fishery in rural pond by bringing, hitherto unused, land resources in to productive cycle.

Waterlogged Areas

Water logging is a hydro geological condition, when ground water table reaches shallow critical limits of root zones where soil profiles are saturated with capillary water. There are four categories of water logged areas. 1. Pounded/flooded: water get accumulated temporarily in depression zones and get increased due to escape from canals. 2. Marshy land: here most of the year ground water level is above the ground surface and possess hardy weeds. 3. Critical water logged areas: here ground water is in critical condition and reaches the surface with capillary rise. Generally it varies from 1.5 m to 3.00 m below ground level. 4. Semi critical water logged areas: here more in monsoon season but capillary water do not always reach surface. Most of the degraded wastelands/wetlands are the consequent effect of water logging. In Uttar Pradesh the waste land were estimated up to 40.88 lac ha in 1991-92. The extent of degraded lands is high in central Gangetic plains (17.36 per cent) and eastern plains (12.37 per cent and poor in western plains (10.265). Water logging is responsible for secondary usarisation, concretion of soil profile, drying of large trees, ground water pollution, agriculture production loss and health hazards. Scientific management of these areas to reverse the degradation and full utilization is possible through judicious exploration of its uses

Figure 14.3: Weed Infested Waterlogged Area

UP Water Sector Restructuring Project (UPWSRP)

This is World Bank funded project under implementation in Gomti and Ghaghra basins to develop sustainable farm sector activities through optimum utilization of water resource. Total cost of the project is Rs. 6000 crores for 12 to 15 years duration. Although project is formulated for development of irrigation facilities but fisheries is also a component especially for water logged areas. Due priority to this aspect will certainly give credence about the mitigation of water related problems through productive utilization of waste resources *i.e.* fishery.

Alkaline/Saline Soil

Vast land area belongs to this category and remains unutilized in our country. Considerable efforts have been made to reclaime these areas by physical and chemical treatment. Shift in strategy by including fisheries from beginning will help in creation or renovation of new avenues for production.

Sodic Land Reclamation Project

Reclamation of Sodic land was initiated in Uttar Pradesh during 1979-80 through corporation with share capital of Rs. 100 lac but desired progress could not be achieved in absence of fund shortage. State government later on clubbed this task with other on going schemes of agriculture and rural development. However major impetus to this activity was with the approval of two separate projects one from World Bank and other from European Economic Community. One of the important tasks of these projects is assured water management and creation of drainage network. Incorporation of fisheries with this reclamation work will not incur any significant financial burden but certainly result in to creation of new assets suitable for production.

Under Groundwater

Availability of groundwater plays a crucial role in fish culture during summers. Farmers can supplement water to extend culture period in non perennial ponds. Storage of water directly helps in recharging of aquifers hence both these water sources are complementary to each other. In Rajasthan, Tarun Bharat Sangh has restored aquifers by constructing 8,600 tanks, ponds and check dams before that there was no such efforts which caused irreparable loss to agriculture and local employment. This activity restored the water table to such an extent that the affected areas have today been declared white zone (no water shortage), agriculture is flourishing again, and local livelihoods are booming (Aiyar 2006).

Rain Water Harvesting

Most of water precipitation in the form of rain goes in to drains from cities due to impermeable surface. It do not percolate fully to supplement under ground storage in these areas which are already more consumptive due to dense human habitation. Construction of water harvesting structures with sufficient open surface retention would be helpful in restoring aquifers as well as eco-stabilizers.

Check Dams for Irrigation and Water Recharge

Plateau region of UP is water deficient area and depends on rain water for irrigation as well as ground water recharge. Construction of check dams on smaller rivulets/drains creates water storage sufficient to irrigate about 20 ha area. These water bodies require little modification in design and construction to make suitable for fishery also. About 1013 such structures are already there and more are coming up in that region.

Restoration of Traditional Water Bodies

This scheme covers renovation of about 724 rural ponds in various dark zones of UP by minor irrigation department to harness rain water for community use and recharge purpose with a total cost of about 27.29 crores. The dead storage height in these ponds is only one meter which becomes insufficient for fish culture. This huge investment can be of more productive use if include fishery as income generating activity in these ponds.

Information Technology

Modern advances in information and data processing technologies have tremendously increased the capacity of humans to analyze the complex multiple resource use options and to network large number of people in to integrated decision making structures. To harness common property resources, application of this modern tool is essential for the creation of data base like GIS, MIS and other information for effective linkages between the center and state, among public-private- NGO-CSO systems and among research-education–extension-farmer-market. In the present case, it is much more relevant because of wide distribution of water bodies in remote areas, fluctuation in water spread area and multiple uses of related resources.

Convergence and Synergetic Use of Resources

The present trend of land and water development with a single objective of increasing agriculture production has led to the neglecting of other related sectors. In the past the water development included intervention aimed at flood protection, drain wet lands and provision of irrigation for agriculture by pumping surface or under ground water. These have led to loss of habitat and decline in open water fishery. It is now realized that great opportunities exists for utilization of land and water resources of the country in an integrated, balanced and comprehensive manner that will not only lead to growth of various farm activities but also resolve many environmental problems. The concept of unitary culture of fish, crop or animal husbandry has gradually been changed to the concept of integrated system with a view of augmenting production through the development of structural network in line with local conditions, comprehensive utilization of various farm products and recycling of wastes. Integrated fish farming is a multiple land use approach in fish culture which combines fish farming with other agriculture production. The benefits derived from judicious integration of various resources with fish farming include better utilization of land and water resources, effective utilization of wastes and byproducts, drastic minimization of operational expanses often incurred through the use of feeds and fertilizers, ultimate improvement in the living conditions of the rural community due to increased employment and income generation and maintenance and management of ecologically balanced ecosystems.

Acknowledgements

The author is thankful to director fisheries, GOUP, for constant encouragement and valued suggestions during this work. Thanks are also due to Mr. Ali Anas for computer work.

References

Abdul Kalam, A.P.J. and Y.S. Sunder Rajan, 2004. *India 2020: A Vision for the New Millennium.* Penguin Books India (P) Ltd., New Delhi, p 191.

Anon, 2006. *Souvenir: All India Seminar on Development of Minor Irrigation Works and its Impact on Agriculture Production.* Minor Irrigation Department, UP. January 16–17, Lucknow.

Anon, 2005. *Handbook of Fisheries Statistics* 2004. Department of Animal Husbandry and Fisheries, Ministry of Agriculture, GOI, New Delhi.

Anon, 2003. *Irrigation-vision 2025.* Irrigation Department, Government of UP, Lucknow.

Anon, 2004. *Agriculture Statistics of Uttar Pradesh.* Publication No. 246. Agriculture Department, UP.

Aiyar, S.S.A., 2005. *Convert Canals to Fisheries.* Times of India, Lucknow, November.

Aiyar, S.S.A., 2006. *Livelihoods: The Man in the Streets*. Times of India, Lucknow. January 22.

Krishna Rao, D.S., 2005. Strategies for enhancement of fish production from Indian reservoirs. *Fishing Chimes*, 25(1): 86–90.

Mishra, A. and Mohanty, R.K., 2004. Productivity enhancement through rice-fish farming using a two-stage rainwater conservation technique, 67(2): 119–131.

Nath, D. and A.K. Das, 2004. Present status and opportunities for the development of Indian riverine, reservoir and beel fisheries. *Fishing Chimes*, 24(1): 61–67.

Padey, M., 2004. Uttar Pradesh agriculture digests: 2001–02. UP Agriculture Statistics and Crop Insurance Directorate, Krishi Bhawan, Lucknow, India.

Singh, R.B., 2005. Agricultural vision. *Agriculture Today*, February, p 60.

Sultan, S., 2004. Wetland fishery management in Uttar Pradesh. *Third Indian Fisheries Science Congress.* November 4–6, New Delhi, p. 127.

Sundaram, K.V., 2006. Sustainable management of land resources: The challenges before current millennium. Proceedings national seminar on sustainable management of land resources through land care movement. *National Seminar on Sustainable Management of Land Resource s through Land Care Movement*. Yojna Bhawan, Lucknow. January 12–13.

Vass, K.K., 2005. Future challenges of capture fisheries in India. *Fishing Chimes*, 25(1): 60–64.

Chapter 15

Development and Management of Reservoirs in India through Morpho-edaphic Index and Stocking Density: A Review

B.R. Kiran, E.T. Puttaiah and K. Harish Kumar

Introduction

Reservoirs are primarily constructed for irrigation and power generation besides drinking water and fisheries is incidental. According to Sugunan (1995) the area of reservoirs in India is about 3 million ha. The average fish production from Indian reservoir is less than 10 kgha^{-1} yr^{-1} (Jhingran, 1989). It is necessary to know their potential productivity for suitable management programmes of fisheries of reservoirs. Sreenivasan (1966) had done pioneering work on reservoir limnology and fisheries from impoundments. Karnataka has a large number of minor and major tanks (area : 213404 ha) and reservoirs (74 ; area : 223887 ha ; Sugunan, 1995).Reservoirs constitute the prime inland fishery resource of India both in terms of surface area and production potential. More than 3 million ha of man made lake in the country after immense scope for fish yield optimizations through capture fisheries, culture-based fisheries and various other forms of enhancement. In India, reservoirs are classified as small (< 1000 ha) medium (1000-5000ha) and large (> 5000 ha) (Sugunan, 1997). Nearly half of the surface area of reservoirs in India comprises small reservoirs, which are by and large managed as culture based fisheries. Medium and large reservoir s spread over more than 1.5 million ha, where a combination of culture and capture based fisheries is practiced. Inspite of fish production potential of these categories of reservoirs no attempt has been made to construct their fish yield prediction models based on morpho edaphic index (MEI) there is a need to estimate the stocking rate of these reservoirs based on the maximum sustainable yield. It is relatively easier to estimate the stocking density in culture based fisheries as the management is based on a stocking and recapture basis. Stocking and other management strategies to be followed in small reservoirs of India have been described by many workers (Jhingran and Sugunan, 1990; Sugunan, 1997) these includes methods for calculating the stocking density on the basis of individual growth rate of stocked fish and expected fish yield of the waterbody (Welcomme, 1976; Sugunan and Sinha, 2000). However, such formulae have little applicability to medium and large reservoirs, where natural populations contribute a considerable part of the catch. In larger water bodies the main objective of management is still on exploitation of neutralized populations stocking, wherever restored to be, done with a view augmenting the yield. (Owx,1994) and Welcomme (1996) emphasized the need for developing stocking strategy for reservoirs and De Silva (2001) stressed the importance of stocking based on the potential yield, predicted through use of appropriate empirical models.

Ever since Rawson (1952) suggested that the fish yield potential is related to the mean depth of inland lakes and reservoirs, many authors have tried to assess fish yield potential from a number of physico chemical and morphometric parameters (Northcote and Larkin, 1956; Haynes, 1957). However, the models were cumbersome involving too many physico chemical and morphometric parameters. Ryder (1965) developed morpho edaphic index (MEI) by taking into account total dissolved solids and mean depth to represent respective edaphic and morphometric parameters of the lakes. This tool has since been used to determine potential yield from reservoirs in Canada (Matuszak, 1978), Africa (Henderson and Welcomme, 1974), Sri Lanka (Wijayarathne and Amarasinghe, 1984; Nissanke *et al.*, 2000) and Bangladesh (Hansan *et al.*, 2001). An (2001) reported model to predict yield based on chlorophyll and specific conductivity. Estimation of potential fish production made through morpho edaphic index can be utilized for calculating the stocking densities of fish fingerlings in inland reservoirs (Welcomme, 1976) workers on North American lakes (Young and Heimbuch, 1982) have opined that surface area can alone be the predictor of fish yield. But this model based on area cannot be considered useful to estimate yield from reservoir as both edaphic and morphometric parameters need to be used as the basis for prediction (Ryder *et al.*, 1994). At present most of the Indian reservoirs are overexploited and there is scope for enhancing the fish production by stocking the reservoirs.Even though more than 100 reservoirs in India have been subjected to scientific investigations most of them are limited to limnological and faunistic survey, with very little emphasis on population dynamics (Sugunan, 2000). In order to develop guidelines for estimating stocking densities in medium and large reservoirs, it is essential to estimate the fish production potential. An attempt has been made to establish a relationship between MEI and MSY of all species and also relationship between MEI and present yield from reservoirs in India.The relationship between yield and other characters like area, depth and specific conductivity has been worked out. Further, a method of estimating stocking requirement of reservoirs has been attempted.

Methods

Time series data on catch and effort as specific conductivity, area and mean depth for seven reservoirs were collected from published literature. Maximum sustainable yield was calculated using Schaefer" model (Sparre *et al.*, 1989) Data on area, depth specific conductivity and yield on seventeen reservoirs were collected from published literature in order to examine the relationship between yield and MEI. Morpho edaphic index was calculated by dividing specific conductivity (mmhos cm^{-1}) by mean depth in meter (Henderson and Welcomme, 1974). The relation between yield and other character was worked out and regression lines were fitted using standard statistical tools (Snedecor and Cochran, 1967) and MS excel computer software.

In the present study, two models have been employed to arrive at an estimate of potential yield *viz.*, trophometric model of Melack (1976) and 0.2 per cent of gross primary production as potential yield. Primary production has been estimated using dark and light bottle method (Vollen Weider, 1969).

Limnology and Fish Production Efficiencies of Reservoirs

In this study coarse and fine sand predominate the basin soil in all the reservoirs except in Narayanpur. In the catchment areas of reservoirs in the Western Ghats, which consists of red soil, with low erosion, the basin texture predominantly remained sandy.

Sub surface maxima in primary production occurred in most of the reservoirs during both pre and post monsoon months which reveals photo inhibition. The depth profiles were mostly characterized by low production per unit volume and absence of marked maxima during all the seasons. With the exceptions of Hemavathy, Nugu and Manchanbele, maximum photosynthetic efficiency (A_{max}) values were higher in the impounds during pre monsoon with Narayanpur recording the highest (178 mg Cm^{-3} h^{-1}) and the lowest (20 mg Cm^{-3} h^{-1}) in Ghataprabha.

The high organic load from the catchments received by Harangi is reflected in the rich soil organic carbon (2.7 per cent) and available nitrogen (62 mg–100^{-1} g). This reservoir could be placed under productive category. The yield potential of the reservoirs has been estimated (Rama Krishnaiah *et al.*, 2000) to be between 44 and 104 kg ha^{-1} (Table 15.1).

Table 15.1: Fish Yield Potential (kgha^{-1}) of Karnataka Reservoirs by Different Models (Sukumaran and Das 2005)

Reservoirs	*Melack's Model*	*0.2% of Gross Production*	*Achieved Targets (kgha^{-1})*
Bhadra	34	80	30–40
Linganamakki	42	99	30–40
Harangi	44	104	60–75
Kabini	34	80	50–70
Nugu	38	92	80–100
Hemavathy	36	85	50–60
Manchanbele	85	168	170–180
Vani Vilas Sagar	32	72	60–70
Narayanpur	59	131	100–130
Ghataprabha	25	47	30–40
Malaprabha	27	58	30–40

Considering all the productive features of Kabini reservoir the yield predicted is 34 to 80 kg ha^{-1}. Inspite of strong oxycline during summer and post monsoon season Nugu reservoir indicated high productivity; the expected yield is about 38 to 92 kg ha^{-1}. The potential yield of Hemavathy was 36 to 85 kg ha^{-1}, which is considered as low and is due to high turbidity prevailing therein. The yield potential of Manchanbele reservoir is high in the range of 85 to 168 kg ha^{-1}. This may be due to nutrients enrichment in the newly inundated impoundment (Cooper 1966).

Potential Yield (PY)

PY can be predicted, based on the existing models like morpho-edaphic index (MEI) method, tropodynamic model, Gullands model of catch and effort data (Jhingran 1986). Following Melack (1976) a regression equation has been drawn for fifteen Indian reservoirs as described below.

$$\text{Log FY} = 0.122\ \text{PG} + 0.95$$

where,

FY is fish yield and PG is gross photosynthesis. However, this has been shown to predict considerably low fish yields.

The prediction of potential fish yield from lakes and reservoirs is very important for setting valid targets for fisheries development in such waterbodies.

Stocking

Stocking is virtually the mainstay of fisheries management in reservoirs and lakes and therefore needs top priority. The stocking rate can be calculated based on the average growth rate of individual fish and the expected production utilizing the formula (Huet 1960).

$$\underset{(\text{No. of fingerlings ha}^{-1})}{\text{Stocking rate}} = \frac{\text{Expected production (kg)}}{\text{Average individual growth rate(kg)}} + \text{Loss due to mortality or escapement (\%)}$$

Considering the need for rapid assessment of country reservoir resources, the following planning norms are suggested.

1. An inventory of all the reservoirs large (> 100 ha) medium (500 -1000 ha) and small (< 500 ha) in the state may be prepared along with the estimates of their fishery potential, calculated as per guidelines given above. The reservoirs may then be divided into (a) reservoirs, which can be developed on stocking cum, capture fisheries principles having significantly high potential yield (b) reservoirs, which can be developed on extensive aquaculture techniques as put and take system and also have significant potential yield (c) reservoirs intermediate in size and potential yield.
2. The reservoirs in the state may then be sub classified in terms of priority based on their productive potential and necessary infrastructure requirement vis a vis resources availability. Selected reservoirs may be taken up for maximizing fish production from them. Such an approach has been suggested in some tropical countries by Welcomme and Henderson (1976).

Maximum Sustainable Yield (MSY) and Morpho-edaphic Index (MEI)

The values of area, MSY and MEI for seven reservoirs are presented in Table 15.2. The relation between MSY and MEI has been found to be non linear and significant. It can be described by the equation:

$$Y = 0.9897 * X^{1.3888} (R^2 = 0.7366) \quad .....(1)$$

where,

Y represents MSY and X MEI (Figure 15.1.)

Table 15.2: Details of Reservoirs in Respect of Area Mean Depth, Specific Conductivity, MEI and Maximum Sustainable Yield

Reservoirs	*Area (ha)*	*Mean Depth (m)*	*Specific Conductivity*	*MEI*	*MSY (kg ha⁻¹)*	*Sources of Data*
Bhavanisagar	7876	11.9	120	10.08	31.8	Natarajan *et al.*, 1981
Rihand	46538	22.8	92.24	4.04	3.9	Natarajan *et al.*, 1981
Pong	24629	35.7	181.4	5.82	23.4	Sugunan 1995
Tunga Bhadra	37814	9.9	197	19.9	89.6	Singit *et al.*, 1987
Stanley	15346	17.24	281.2	16.31	23.3	Sreenivasan 1976
Gumti	4500	5.236	87.1	16.63	50	Bhattacharya and Saha 1986; Sugunan, 1995
Govindsagar	16867	11.9	195	16.39	46.13	Anon, 1984

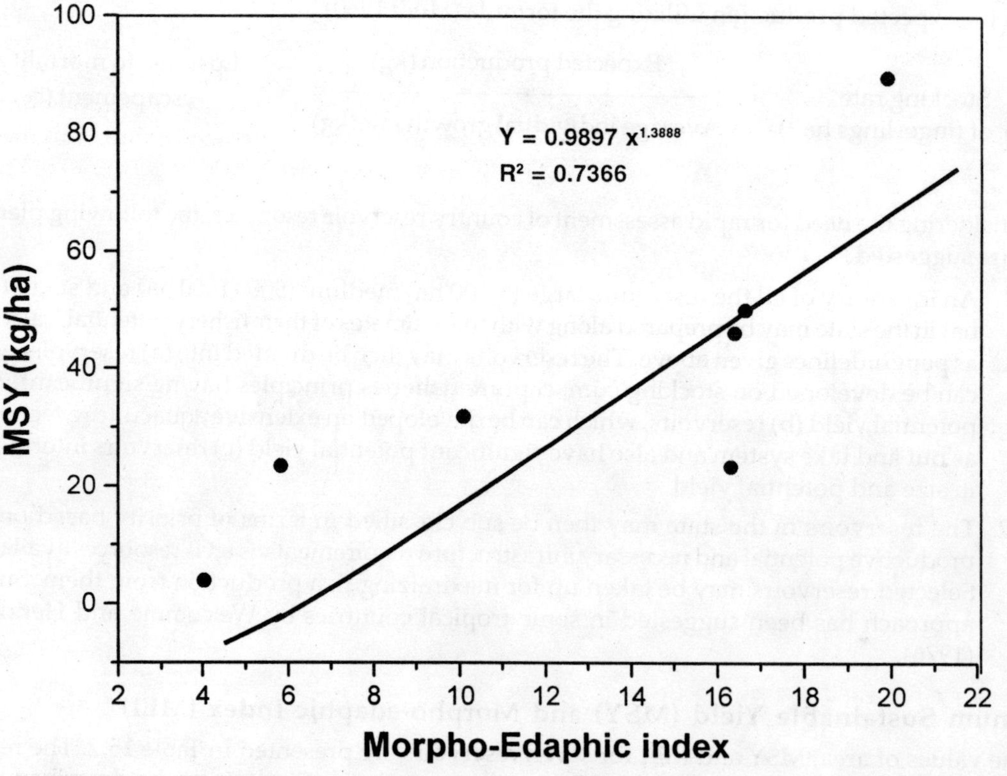

Figure 15.1: Relationship Between Maximum Sustainable Yield and Morpho-edaphic Index (Sugunan *et al.*, 2002).

The details of seventeen reservoirs in respect of area, yield (mean) and MEI are presented in Tables 15.3 and 15. 4. The relation between depth and yield has been found to be non-significant. The relationship between yield and area of waterbody is non-linear, significant ($y=2103\ x^{-0.509}$, $R^2=0.442$) and negative indicating decrease of yield with the increase of the size of water body. This is in agreement with the findings in Chinese and Mexican reservoirs (Welcomme and Bartley 1988 a and b). However, the relationship between yield and specific conductivity is positive (Figure 15.2) which indicates that yield increases with the increase of specific conductivity.

Table 15.3: Source of Secondary Data

Name of the Reservoirs	*Source of Data*
Hemavathi	Devaraj *et al.*, 1987
Kabini	Murthy *et al.*, 1986
Mettur	Sreenivasan 1992
Ravishankar sagar	Desai and Soni 1994
Nagarjun sagar	Natarajan and Pathak 1983
Tawa	Srivastava *et al.*, 2000
Pong	Ramakrishnaiah and Govind 1993
Tilaiya	Ramakrishnaiah and Govind 1993
Hirakud	Sugunan, 1995
Marconahalli	Ramakrishnaiah *et al.*, 1998
Vaigai	Sreenivasan 1992
Sathanur	Sreenivasan 1992
Morel	Kaushal and Sharma 2001
Guda	Kaushal and Sharma 2001
Panchana	Kaushal and Sharma 2001
Baretha	Kaushal and Sharma 2001
Ramgarh	Jhingran 1989

The relationship in case of area and specific conductivity has also been found to be significant at 1 per cent level. But the models based on area or specific conductivity is seldom used for estimation of yield. The concept of morpho-edaphic index has been evolved over the years by different workers. The different workers, who established relation among abiotic and biotic variables in reservoir ecosystem. Essentially, these formulae strived to incorporate all the variables (both morpho metric as well as edaphic) having a bearing on productivity. A global formula should be simple with minimum number of variable in order to enable the managers to use it with ease. Recognizing this, mean depth and total dissolved solids were taken as two key variables, which can represent morphometric and edaphic characters of the lake respectively (Ryder *et al.*, 1974). Since Ryder (1965), MEI has been taken as the standard tool for predicting yield.

The relationship between mean yield (Y) and morpho-edaphic index (MEI) has been found to be non-linear and significant and presented below and shown in Figure 15.3.

$$Y = 3.984^{*}\ X^{0.6374}\ (R^2 = 0.4681) \qquad ....(2)$$

This formula can be used for calculating the mean yield of Indian reservoirs

It is apparent from Table 15.3 that the present level of production as estimated by the relationship between yield and morpho-edaphic index is 33,7 kg, whereas the MSY is 169.27 kg, thus an average increase of 135.58 kg, may be obtained from the reservoirs. This can be achieved by stocking the reservoirs with fingerlings based on the following equation:

$$\frac{MSY_{est} - Y_{est^{*}}}{W} \qquad S = e^{z}\,(t_i - t_0) \qquad .....(3)$$

S = Number of fish fingerlings to be stocked in addition to the present recruitment to achieve the estimated value of maximum sustainable yield, Z = total mortality rate (Welcomme, 1976), t_i = age at

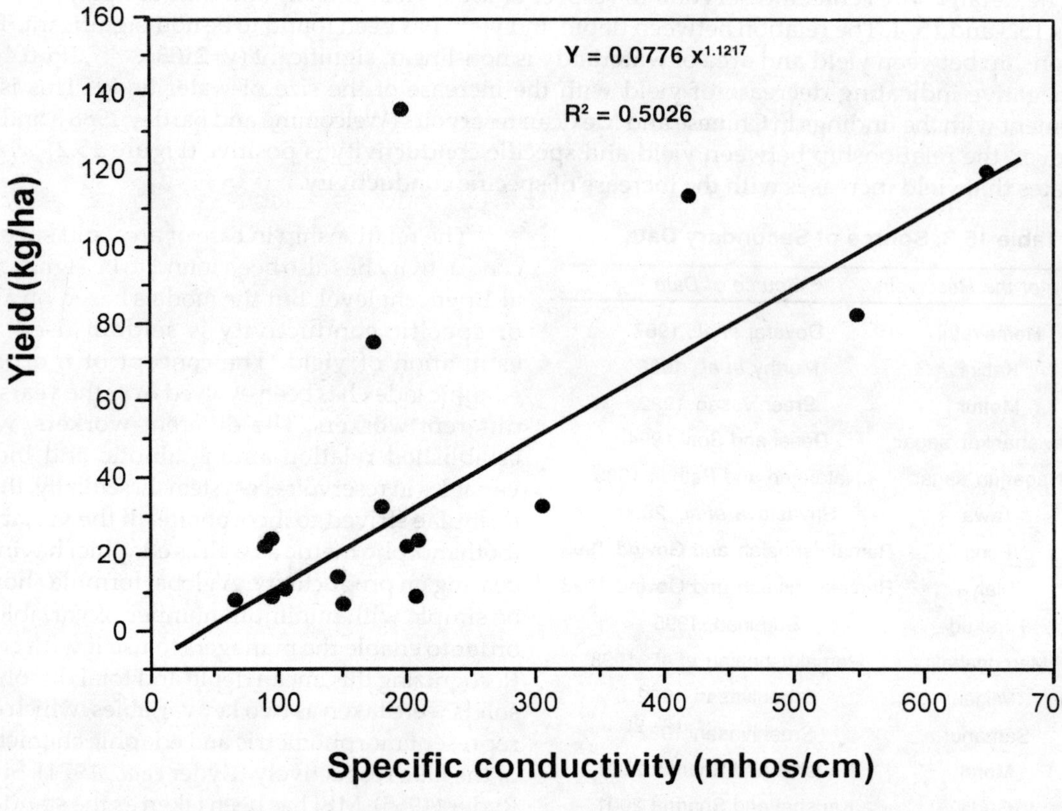

Figure 15.2: Relationship Between Yield and Specific Conductivity of Indian Reservoirs (Sugunan *et al.*, 2002)

capture, t_0 = age at recruitment, W = mean weight of fish at capture, MSY_{est} = maximum sustainable yield estimate using morpho-edaphic index relationship (formula 1) and Y_{est} = Present mean yield estimated by using morpho-edaphic index (Formula 2).

The estimated number to be stocked has been shown in Table 15.3 based on the following assumptions

1. The average weight of fish harvested from medium and large reservoirs is 500gm and
2. The total mortality (z) of the stocked fingerlings is -1.00, which has been worked out on the basis of Welcomme (1976) taking survival as 37 per cent.

Ryder *et al.* (1974) and Henderson and Welcomme (1974) have discussed the weaknesses and strengths of the yield prediction models involving MEI. In India, many workers have expressed reservations (Natarajan,1979; Sreenivasan, 1989) about the use of MEI in the Indian context. The main drawbacks pointed out by them are the wide fluctuation in the discharge rates. However, this problem is common to man made lakes everywhere. Welcomme (1976) expressed the view that MEI can be used to estimate yield in the lakes where the volume and temperature are fairly constant through out the year and the ionic composition of water is dominated by carbonate and bicarbonate ions. While, temperature variation in Indian reservoirs is within a relatively narrow range (Sugunan,1995) compared to their counterparts in temperate countries, the condition about carbonate and bicarbonate ions is adequately met. Model developed by Henderson and Welcomme (1974) takes into consideration 19 African lakes, which have similar heterogeneous components. Reservoir of Sri Lanka shares comparable morpho-edaphic profiles with those of peninsular India and the MEI models developed for Sri Lankan reservoirs are reported to be working with satisfactory results (Mendis,1965; Wijayarathne and Amarsinge,1984). Marshal (1984) examined the relevance of morpho-edaphic index and its components as attributes to predict yield in African reservoirs. Notwithstanding the various draw backs, which are inherent to the system of creating models based on a limited number of parameters, some forms of yield prediction models are necessary to initiate management norms for reservoirs in India. Therefore, the model suggested in this paper can be used for making a first approximation of fish yield potential of reservoirs in India.

Table 15.4: Area, Mean Depth, Specific Conductivity, MEI, Estimated MSY, Estimated Yield and Estimated Number to be Stocked in Indian Reservoirs (Sugunan *et al.*, 2002)

Reservoir	*Area (ha)*	*Mean Depth (m)*	*Yield (kg ha^{-1})*	*Sp. Conductivity (mmhos cm^{-1})*	*MEI*	*Yield (kg ha^{-1})*	*MSY (est) (kg ha^{-1})*	*Yield (est) (kg ha^{-1})*	*MSY (est) (kg ha^{-1})*	*Expected Enhancement in Production (t)*	*Number to be Stocked*
Hemavathi	8502	12.4	14	147	11.85	14	30.67	19.3	11.41	97	71347
Kabini	6020	9.1	22.0	90	9.89	22.0	23.86	17.2	6.69	40.3	29642
Mettur	14690	17	23.6	210	12.23	23.6	32.04	19.7	12.39	182	133898
Ravishankarsagar	9540	9.5	8.0	67.2	7.07	8.0	14.97	13.9	1.11	10.6	7786
Nagarjunasagar	28475	40.5	8.8	95.5	2.36	8.8	3.26	6.9	-3.63	-	-
Tawa	20055	11.0	24	95	8.64	24	19.78	15.7	4.03	80.8	59418
Pong	24529	35.7	32.3	181.4	5.82	32.3	11.42	12.2	-0.82	-	-
Tilaiya	5921	6.6	10.8	105.50	15.98	10.8	46.45	23.3	23.15	137.1	100844
Hirakud	71963	11.3	6.9	151	13. 36	6.9	36.23	20.8	15.43	110.7	817174
Marconahalli	1337	5.1	75.0	175	34.31	75	134.24	37.9	96.31	128.8	94744
Vaigai	2420	8.1	8.9	208	25.67	8.9	89.73	31.5	58.20	140.8	103623
Sathanur	2420	10.0	113.1	420	42	113.1	177.78	43.1	134.63	325.8	239706
Morel	1564	4.9	119.4	649	132.45	119.4	876.21	89.7	786.49	1230.1	905032
Guda	1859	5.1	32.5	306	60	32.5	291.74	54.2	237.58	441.7	324954
Panchana	1240	4.8	22.0	201	41.87	22	177.01	43.1	133.95	166.1	122206
Baretha	1006	5.2	135.6	197	37.88	135.6	154.03	40.0	113.63	114.3	84103
Ramgarh	1260	4.6	82.1	549	119.35	82.1	758.22	84.0	674.26	849.6	625073
Average							169.27	33.7	135.58		

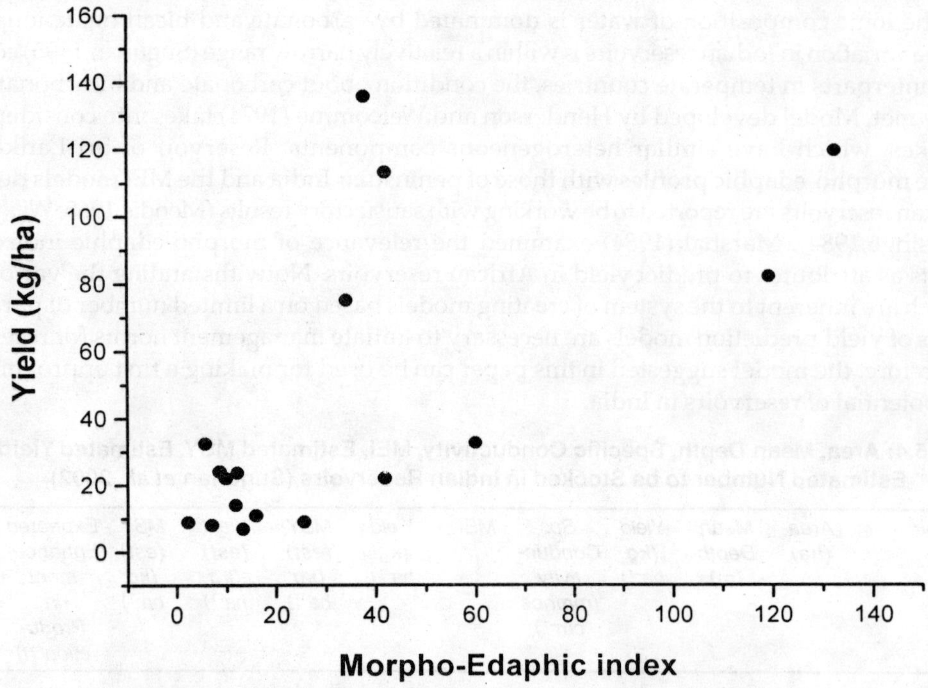

Figure 15.3: Relationship Between Yield and Morpho-edaphic Index (Sugunan *et al.*, 2002)

The significance of the regression suggests the validity of the models. Stocking density in general are calculated by dividing the yield rate with individual growth before adding some allowances to loss due to mortality (Huet,1960;Welcomme,1976). However, Welcomme pointed out that such models are not applicable in reservoirs where the natural recruitment has a significant contribution. Moreover, unlike the culture based fisheries of small reservoirs, where the yield is dependent totally on stocked fish, in medium and large reservoirs, stocking is done to augment the fish production, thus necessitating stocking strategy based on MSY. Therefore, the expected yield rate 'P' (in the numerator of Welcomme's formula) has been replaced with $MSY_{est-} Y_{est}$ in order to avoid possible over stocking.

The model presented in this paper has been developed on the basis of MSY of all species. However, in multispecies fisheries, it is possible to estimate the stocking requirement of each species if MSY is known separately for the species (Welcomme, 1976). India has large area of small, medium and large reservoirs which need to developed on the basis of culture based fisheries. At present, no scientific advice is available to estimate the stocking rate. The formula (3) is an attempt to provide scientific guidelines for stocking reservoirs. Although not very accurate, it can be considered as a first attempt to develop such stocking guidelines.

Equation (1) presented in this article can be tentatively used for calculating the maximum sustainable yield of Indian reservoirs and equation (2) can be used for estimating the present mean yield based on morpho-edaphic index. Methods described by Huet (1960) and Welcomme (1976) are adequate to calculate the stocking density of culture based fisheries of small reservoirs in India. The stocking rate for medium and large reservoirs can similarly be estimated, using the equation (3) shown in this article.

The fish production potential from 4 medium reservoirs at 1 per cent Gross primary production (GPP) ranged from 108.3 to 767.3 kg ha^{-1}. But the actual fish yield from the reservoirs has been low (3.9–33.8 kg ha^{-1}) in spite of a reported stocking of fish seed at the rate of 140 to 450 numbers per hectare per year. The stocked varieties of fish are not adequately reflected in the catch and their contribution to the total fish landings are meager (1.2–3.4 kg ha^{-1}). The conversion efficiency from GPP to fish yield in these four reservoirs (Tamil Nadu) has been poor, ranging from 0.001 to 0.05 per cent.

At present none of these reservoirs are scientifically managed. The size of seed at stocking has been very small (15-60 mm). The poor recovery (1.2-3.4 kg ha^{-1}) of stocked fish indicated large-scale mortality due to various reasons such as predation, stress and strain of long distance transportation of fish seed before stocking, unhealthy seed succumbing to various diseases, natural mortality, escapement, etc. In the absence of natural recruitment and auto-stocking of commercially important fishes in these medium reservoirs, stocking from external sources hold important role in improvement of the fishery.

Murugesan *et al.* (1999) and Selvaraj *et al.* (2000) recommended various management measures for enhancement of fish production from small reservoirs. Some of the recommendations therein can also be applied to medium reservoirs. Fish Seed Committee (1976) suggested a stocking rate of 500 seed ha^{-1} whereas, Selvaraj *et al.* (2000) recommended 300 to 400 advance fingerlings ha^{-1} for small reservoirs. A stocking density of 200-250 advanced fingerlings (100-150mm length) is recommended for the medium reservoirs. The self propagating fishes such as wed fishes, minor carp and predatory fishes, which have low market value dominated in the fish catches from these reservoirs, indicating that the bulk of the food resources of the ecosystem have been consumed by them and thereby they have suppressed the establishment of stocked varieties of fish. For reducing the competition from trash fishes for food and shelter with cultivable carps and to encourage the latter, it is suggested that the low value fishes must be systematically fished out and that high values fishes with fast growth and food conversion efficiency must be stocked immediately after accumulation of rain water to occupy all the niches of the system.

Stocking of healthy fingerlings at appropriate species combination according to the food availability in the reservoir ecosystem at optimum density will certainly improve the fish production as it happened in Aliyar and Thirumoorthy reservoirs (Selvaraj *et al.*, 2000). Judicious exploitation with adequate fishing gears will result in optimum sustained yield from these reservoirs. As the fish yield obtained from reservoirs in Tamil Nadu are far below the production potential assessed through primary production, there is great scope for enhancement of fish yield through sound management programme.

Changing the environmental variables to enhance the primary production is difficult in large aquatic system. However, scientific management to the biological advantage of fish could manipulate biotic communities. Because of the divergent geo-climatic, morphometric and edaphic environment under which the reservoirs are situated, it is difficult to evolve common management practices for them. Systematic and integrated study is warranted in the various reservoirs in Karnataka to establish criteria for planning their fisheries development It is felt that under the prevailing socio-economic conditions, such shortage range studies undertaken for reservoirs would provide rapid assessment of their fisheries potential to take up fish culture.

Management Intervention through Stocking

Stocking of fish seed in small reservoirs of India has been found to be a useful tool for developing their fisheries potential. Stocking of economically important fast growing fishes from outside is aimed at colonizing all the diverse niches of the biotope for harvesting maximum sustainable fish from them. As estimated, 500 fish fingerlings were stocked in the small reservoirs. The stocking ratio was maintained as Catla 40 per cent, Rohu 30 per cent, Mrigal 20 per cent and Common carp 10 per cent. There is a direct correlation between the stocking rate and catch per unit effort (Jhingran *et al.*, 1981; Sugunan 1997). Sugunan (1995) reported that effective stocking policies raised the production from 1.67 kg ha^{-1} in 1964 to 194 kg ha^{-1} in 1985 a reservoir, while in other, the yield rose from 10 kg ha^{-1} in 1966 to 107 kg ha^{-1} in 1991. Hence, 'stocking' can be considered as the most important management tool for increasing fish yield from small reservoirs.

References

An, N.Q., 2001. Effectiveness of stocking in reservoirs in Vietnam. In: *Reservoir and Culture Based Fisheries: Biology and Management*, (Ed.) De Silva S.S. Australian centre for International Agricultural Research, Canberra, 235–245.

Anon, 1984. *Proceedings of the VII workshop on All India Coordinated Projects on Reservoirs Fisheries*, Patna, 11–12 July, 117pp.

Bhattacharya, T. and Saha, R.K., 1986. A comparative study of the physico chemical conditions of the water and plankton of Gumti Reservoir, open channel and river below Hydel project in Tripura. *Proc. Nat Symp. Fish and Env.*, p. 61–63.

Cooper, G.P., 1966. Fish production in impoundments. *Re. Mich. Dep. Conserve. Res. Dev.*, 104, p. 12.

Cowx, I.G., 1994. Stocking strategies. *Fish Manag. Ecol.*, 1: 15–30.

De Silva, S.S., 2001. Reservoir and culture based fisheries: Biology and management. Proceedings of an International workshop, Bangkok, Thailand, 15–18 February 2000. *ACIAR Proceedings* No. 98, 384 pp.

Desai, V.R. and Soni, S.L., 1994. Potential fish yield of Ravishankarasagar Reservoir (Dist. Raipur, M.P.) Through Chemical parameters of water. *J. Inland Fish Soc., India,* 26: 1–8.

Devaraj, K.V., Gopalagowda, P., Krishnamurthy, N.M., Balaramagowda, A.R., Mahadeva, H.S. and Fazal, A.A., 1987. Fish and fisheries of the Hemavathy reservoir. *Uni. Agri. Sci.*, Bangalore, p. 83.

Hasan, M.R., Mandal, M.W., Miah, M.I. and Kibria, M.G., 2001. Water quality study of some selected Oxbow lakes with special emphasis on chlorophyll a. In: *Reservoir and Culture Based Fisheries: Biology and Management*, (Ed.) S.S. De Silva, p. 384.

Hayes, F.R., 1957. On the variation in bottom fauna and fish yield in relation to trophic level and lake dimensions. *J. Fish. Res. Bd.*, Canada, 14: 1–32.

Henderson, H.F. and Welcomme, R.L., 1974. The relationship of yield to morpho–edaphic index and number of fishermen in African inland waters. *CIFA Occ. Pap.*, p. 19.

Huet, M., 1960. General Principles of pond construction and management, productivity and stocking. ETAP, FAO Lecture at the II International Training. Bogor, Indonesia 1955, p. 2.

Jenkins, R.M and Morias, D.I., 1971. Reservoir sport fishing effort and harvest in relation to environmental variables. In: *Reservoir Fisheries and Limnology*, (Ed.) G.E. Hall, p. 371–384.

Jhingran, A.G., 1989. Limnology and production biology of two man made lakes in Rajasthan (India), with management strategies for their fish yield optimization. Final report, IDA assisted pilot project on reservoir fisheries management in Rajasthan, Central Inland Fisheries Research Institute, Barrackpore, India, p. 63.

Jhingran, A.G. and Sugunan, V.V., 1990. General guideline and planning criteria for small reservoir fisheries in India. In: *Proceedings of the National Workshop on Reservoir Fisheries*, 3–4 January. Asian Fisheries Society (Indian Branch) Mangalore, p. 1–8.

Jhingran, A.G., Dwivedi R.K., Srivastava K.P. and Singh D.N., 1981. An ecological approach towards stocking policy formulation in Gulariya: A small impoundment. *Symposium on Tropical Ecol.*, Bhopal.

Jhingran, A.G., 1986. *Artificial Recruitment and Fisheries Management of Indian Reservoirs*. Central Inland Fisheries Research Institute Bulletin, 45: 60.

Jhingran, A.G., 1988. *Reservoir Fisheries Management in India*, Bull No. 44, Central Inland Capture Fisheries Research Institute, Barrackpore, India, 68 pp.

Jhingran, 1969. *Methodology on Reservoir Fisheries Investigations in India*. Bull. Central Inland Fisheries Research Institute, 12, CIFRI, Barrackpore, p. 109.

Kaushal, D.K. and Sharma, V.K., 2001. *Ecology Based Fisheries Management in Some Reservoirs of Eastern Rajasthan*. Bull. 108. Central Inland Fisheries Research Institute, Barrackpore, India.

Marshal, B.E., 1984. Towards Predicting ecology and fish yields in African reservoirs from pre impoundment physico chemical data. *CIFA Tech. Pap. Doc. Tech. CPCA.*, 12: 36.

Matuszak, J.E., 1978. Empirical prediction of fish yields of large North American lakes. *Trans. Amer. Fish. Soc.*, 107: 385–395.

Melack, J.A., 1976. Primary productivity and fish yields in tropical lakes. *Trans. Amer. Fish. Soc.*, 105: 575–580.

Mendis, A.S., 1965. A preliminary survey on 21 Ceylon lakes: Limnology and fish production potential. *Bull. Fish Res. Stn.*, Ceylon, 18: 7–16.

Murthy, H.S., Gowda, S.S. and Ayyar, V., 1986. Fisheries of Kabini Reservoir in Karnataka, *Fishing Chimes*, 6: 36–38.

Murugesan, V.K., Rani Palaniswamy and Manoharan, S., 2003. Fish production potential, actual yield and scope for enhancement of four medium reservoirs in Tamil Nadu. In: *Fisheries Enhancement in Inland Waters: Challenges Ahead,* (Eds.) V.V. Sugunan, G.K. Vinci, P.K. Katiha and M.K. Das. Proceedings of the National Symposium 27–28 April 2002. Inland Fisheries Society of India, Barrackpore, pp. 196–203.

Murugesan, V.K., Selvaraj, C. and Manoharan, S., 1999. Techniques for enhancement of fish production from small reservoirs. In: *Nat. Sem. on Development and Transfer of Fisheries Technology*, College of Fisheries, Tuticorin, Tamil Nadu.

Natarajan, A.V., 1979. Ecosystem approach to the development of fisheries in man made lakes in India. Lecture delivered at summer institute on capture and culture fisheries of man made lake in India, *Central Inland Fisheries Research Institute,* 7 July to 6 August.

Natarajan, A.V. and Pathak, 1983. Pattern of energy flow in fresh water tropical and subtropical impoundments. Bull. 3, *Central Inland Fisheries Research Institute*, Barrackpore, India, 37 pp.

Natarajan, A.V. Tripathi, S.D. Desai, V.R. Singh,R.K. Srivasthava, N.P. and Alagaraja, K., 1982. Final Report. *All India Coordinated Project on Ecology and Fisheries of Fresh Water Reservoirs, Rihand. Res.* Inf. Ser. 2. Central Inland Fisheries research Institute, Barrackpore, India, 68 pp.

Natarajan, A.V. Tripathi, S.D. Gopalkrishnayya, C.H. and Abraham, M., 1981. Final Report. *All India Coordinated projects on ecology and fisheries of fresh water reservoirs, Bhavanisagar. Res.* Inf. Ser. 1, *Central Inland Fisheries Research Institute*, Barrackpore, India, 66 pp.

Nissanke,C., Amarasinge, U.S. and S.S. De Silva, 2000. Yield predictive models for the Sri Lankan reservoir fishery. *Fisheries Management and Ecology*, 7: 425–436.

Northcote, T.G. and Larkin, P.A., 1956. Indices of productivity in British Columbia lakes. *J. Fish. Res. Bd. Canada*, 13: 515–540.

Ramakrishnaih, M. Krishna Rao, D.S. Sukumarna, P.K. and Karthikeyan, 1998. *Ecology and fisheries of Markonahalli Reservoir (Karnataka).* Bull. 78, Central Inland Fisheries Research Institute, Barrackpore.

Ramakrishnaiah, M., Rao, D.S. Sukumaran, P.K. and Das, A.K., 2000. *Ecology and Fish Yield Potential of Selected Reservoirs of Karnataka.* Central Inland Fisheries Society of India. Bulletin, 94 p. 41.

Ramakrishnaiah, M. and Govind, B.V., 1993. Effect of ageing on the morphometry and fish productivity of an impoundment. *J. Inland Fish. Soc., India*, 25: 1–6.

Rawson, D.S., 1952. Mean depth and the fish production of large lakes. *Ecology*, 33: 513–521.

Ryder, R.A., 1965. Morphology as dominant factor in the productivity of large lakes *Trans. Amer. Fish. Soc.*, 94: 214–218.

Selvaraj, C., Murugesan, V.K., Manoharan, S. and Karthikeyan, M., 2000. Fish yield optimization in Thirumoorthy reservoir, Tamil Nadu, *CIFRI, Barrackpore, Bull.*, 95: 42.

Shivastava, N.P., Ramakrishnaih, M. and Das, A.K., 2000. *Ecology and fisheries of Tawa Reservoir (Hoshangabad, Madya Pradesh)*, Bull. 100 Central Inland Fisheries Research Institute, Barrackpore.

Singgit, G.S. Ramamohan, R.P. and Krishnaraju, H.V., 1987. Post impoundment changes and effects of conflicting uses on the fisheries of Tungabhara reservoir, India. In: *Reservoir Fishery Management and Development in Asia*, (Ed.) S.S. De Silva, p. 136–144.

Snedecor, G.W. and Cochran, W.G., 1967. *Statistical Methods*. Oxford and IBH Publishing Co., Calcutta. 593 pp.

Sparre, P. and Venema, S.C., 1989. *Introduction to Tropical Fish Stock Assessment Parts I Manual*. FAO Fisheries Technical Paper, 306: 337.

Sreenivasan, A., 1965. Limnology of tropical impoundments III. Limnology and productivity of Amaravathi reservoir (Madras state), India. *Hydrobiologia*, 26: 501–516.

Sreenivasan, A., 1966. Limnology of tropical impounds I. Hydrobiological features and fish production in Stanley reservoir, Mettur Dam. *Int. Re. Ges. Hydrobiologia*, 51: 295–306.

Sreenivasan, A., 1992. Limnology and fisheries of some south Indian reservoirs. In: *Reservoir Fisheries of Asia*, (Ed.) S.S. De Silva, p. 23–37.

Sreenivasan, A., 1989. Principles of ecology and fisheries management of man made lakes. In: *Conservation and Management of Inland Capture Fisheries Resources of India*, (Eds.) A.G. Jhingran, and V.V Sugunan, Inland Fisheries Society of India, p. 275.

Sugunan, V.V., 1995. *Reservoir Fisheries of India.* FAO Fisheries Technical Paper. 345, FAO, Rome, p. 423.

Sugunan, V.V., 1997. Guidelines for management of small reservoirs in India. *Fishing Chimes*, 17: 23–27.

Sugunan, V.V., 2000. Ecology and fishery management of reservoir in India. *Hydrobiol.*, 430: 121–147.

Sugunan, V.V and Sinha, M., 2000. *Guidelines for Fisheries Management in Small Reservoirs in India.* Bull. 93. Central Inland Fisheries Research Institute, Barrackpore, p. 31.

Sukumaran, P.K. and Dal, A.K. 2005. Limnology and fish production efficiencies of selected reservoir of Karnataka. *Indian J. Fish.*, 52: 47–53.

Vollenweider, R.A., 1969. *A Manual of Methods for Measuring Primary Production in Aquatic Environments.* IBP Handbook, 12, Blackwell Scientific Publications, Oxford and Edinburgh.

Welcomme, R.L. and Henderson, H.F., 1976. *Aspects of the Management of Inland Waters for Fisheries.* FAO. Fish. Tech. Pap. 161, p. 36.

Welcomme, R.L., 1996. Stocking as a technique for enhancement of fisheries. *FAO Aquaculture Newsletter*, p. 148–149.

Welcomme, R.L. and Bartley, D.M., 1998a. *Current Approaches to the Enhancement of Fisheries Management and Ecology*, 5: 351–382.

Welcomme, R.L. and Bartley, D.M., 1998b. An evaluation of present techniques for the enhancement of fisheries. In: *Inland fishery Enhancements*, (Ed.) T. Peter, p. 1–35.

Welcomme, R.L., 1976. Approaches to resource evaluation and management in tropical inland waters. Proceedings of the Indo Pacific Fisheries Council, Colombo, October 1976. *Food and Agriculture Organization of the United Nations*, p. 500.

Wijayarathne, M.J.S. and Costa, H.H., 1981. Stocking rate estimation of Tilapia Mossambica fingerlings for some inland reservoirs of Sri Lanka. *Int. Revne. Ges Hydrobiol.*, 66: 327–333.

Wijayratne, M.J.S. and Amarasinghe, U.S., 1984. Estimation of maximum sustainable fish yield and stocking densities of inland reservoirs of Sri Lanka. *J. NARA*, 31: 65–72.

Youngs, W.D. and Heimbuch, D.G., 1982. Another consideration of the Morpho-edaphic index. *Trans. Amer. Fish. Soc.*, 11: 151–153.

Chapter 16

Preliminary Study on Status of Fisheries in Salal Reservoir and Suggestive Measures for Development

K.S. Charak and F.A. Fayaz

Introduction

Freshwater aquatic resources, such as rivers, streams, lakes, ponds and reservoir possess considerable wealth of indigenous fish. These resources are much noted for their variety, biodiversity and production potential. Among all the fresh water resources, reservoirs occupy the prime importance by virtue of their area and production potential and if developed on scientific lines will provide a relatively low priced, good quality animal protein source besides occupational opportunities.

Development of Fisheries in a particular reservoir directly benefits a large number of people engaged in different activities. But, the fish production from these reservoirs is currently much below to their potential. In Indian reservoirs, the fish yield ranges from a minimum of 2.5 Kg/ha/yr to a maximum of 120 Kg/ha/yr (Jyoti *et al.*, 2004). Some of the reasons for low yield are lack of understanding of reservoir ecology, paucity of indigenous fish fauna, faulty stocking and over exploitation. Sugunan and Sinha (2000) emphasized that the fish productivity is related to morphometric, hydrological and ecological features of particular water because the yield is partly a function of biotic and abiotic factors influencing production process of that ecosystem. For management of these reservoirs, estimate of their potential fish yield is a primary pre-requisite and the stocking of the desired species is so adjusted to approach that potential fish yield. Therefore, for proper management of any reservoir, thorough knowledge of existing fish fauna and the prevailing ecological condition is pre-requisite to obtain suitable yield.

The Salal reservoir, first man made medium reservoir in J and K state is constructed by damming river Chenab-an important tributary of Indus drainage system, at Thanpal (Reasi) at an elevation of

496 mtrs. above MSL. A reservoir is an artificial water body formed by damming a river for the purpose of irrigation, power generation, drinking water supply and flood control. According to the area the reservoirs have been classified into 3 categories small (< 1000 ha), Medium (1000–5000 ha) and Big (> 5000 ha). The Salal reservoir has an area of 1000 ha and falls under medium reservoir with total catchment area 21479 sq. kms. The present study deals with the physico-chemical conditions and indigenous fish fauna of Salal reservoir to facilitate formulation of various measures for judicious exploitation and further enhancement of yield from the reservoir. It has been observed that all the physico-chemical parameters are in the productive range. However, among the indigenous fish fauna it has been observed that the Mahseer-an important sport and food fish is dwindling due to construction of dam, which restrict the upward migration of the fish for breeding. There are two beautiful tributaries namely "Rud and Ans", which were the main breeding grounds for Mahseer; but construction of dam has hampered the migration of Mahseer to these tributaries. Hence effecting the breeding of fish. This fish can certainly be protected by adopting conservation and rehabilitation measures and if need felt Mahseer seed being produced at Mahseer hatchery Anji (Reasi), can be stocked in the different water bodies so as to conserve the migratory Mahseer *'Tor putitora'*.

Description of the Reservoir

The Salal reservoir is constructed by damming river Chenab at Thanpal (Reasi) in District Udhampur within the territorial jurisdiction of Jammu and Kashmir State. It is situated at an elevation of 496 mtrs. above MSL at latitude 30°08′ N and longitude 74°50′ E. The details of morphometric features of the Salal reservoir are furnished in Table 16.1.

Table 16.1: Morphometric Features of Salal Reservoir

1.	Location :	
	(a) State	Jammu and Kashmir
	(b) Division	Jammu
	(c) District	Udhampur
	(d) Latitude	30°.08' N
	(e) Longitude	74°.50' E
2.	Dam Elevation	496 mtrs. above MSL.
3.	River	Chenab
4.	Catchment area	21479 sq. kms.
5.	Max. discharge	11895 cums.
6.	Min. discharge	91 cums
7.	Mean annual run off (at Dam site)	17315 million cu. mtrs.
8.	Shore line	487.68 mtrs.
9.	Reservoir capacity at NRL	284 million cu. mtrs.
10.	Dead storage level	487.40 cu. mtrs.
11.	Full reservoir level	494 cu. mtrs.
12.	Submerged area	935 sq. kms.
13.	Max. length of reservoir	40 kms.
14.	Water spread area	1000 ha
15.	Average Depth	36 mtrs.

Figure 16.1: A Panormic View of Salal Reservoir

Figure 16.2: Another View of Salal Reservoir

Figure 16.3: View Near Dam Site

Materials and Methods

After detailed survey of the reservoir four sampling stations were established, station 1st was established near the project site, station 2nd was established towards Rud Nallah, Station 3rd was established towards Arnas village, whereas the Station 4th was established near the Dam site. Monthly water samples from the fixed sampling stations were collected for period of one year from January 2001 to December 2001 in a five liter polythene container and also in 250 ml glass bottles for the estimation of dissolved oxygen at site. The temperature, pH, carbon dioxide, carbonate and bicarbonate were determined immediately after collection whereas, the other physico-chemical parameters were analyzed in the laboratory within 24 hours of collection according to the methods as described in APHA (1995). Calcium and magnesium contents were analysed by titrometry using EDTA, chloride by Argenotmetry and sulphate by turbidemetric methods (Golterman and Clymo, 1971). The data of the parameters were pooled together for discussion. For the purpose of estimation of phytoplankton one litre of water sample was collected by filtering 100 litres of water through plankton net of bolting silk No. 25 (0.06mm mesh size). Lugol's solution was added to the water for preserving phytoplankton and was kept for sedimentation for 24 hrs. and then the identification was carried out with the help of guidelines given by Needham and Needham (1962) and Pennak (1978). Zooplankton collection was made by filtering 100 litres of water through plankton net and was preserved in 5 per cent formalin solution and identification was carried with the help of guidelines given by Needham and Needham (1962) and Battish (1992). The fish fauna were collected by using gill net, preserved in 10 per cent formaldehyde solution with small abdominal incision, brought to the laboratory and were identified with the help of Standard Indian literature (Malhotra, *et al.*, 1975; Talwar and Jhingran, 1991 and Jayaram, 1999).

Resuls and Discussion

The result of various physico-chemical parameters of reservoir are presented in Table 16.2. The air temperature ranged from 14.0° to 30.5°C. January is the coldest month of the year. From January onward, the day temperature increase progressively and reached to maximum in July. The water temperature ranged from 8.30° to 18.40°C; being maximum in June and minimum in January.

Table 16.2: Physico-chemical Profile of Salal Reservoir

S.No.	*Parameters*	*Range*
1.	Air temperature (°C)	14.00-30.50
2.	Water temperature (°C)	8.30-18.40
3.	pH	7.40-8.50
4.	Dissolved Oxygen (mg/l)	7.20-10.30
5.	Free Carbon Dioxide (mg/l)	0.80-1.45
6.	Carbonate (mg/l)	0.00-12.0
7.	Bicarbonate (mg/l)	99.4-216.50
8.	Chloride (mg/l)	0.98-5.93
9.	Calcium (mg/l)	14.30-22.80
10.	Magnesium (mg/l)	2.40-9.96
11.	Sulphate (mg/l)	11.62-34.56

The pH was in the range of 7.4 to 8.5. The minimum and maximum values of pH were recorded in the months of July and April respectively. This pH range clearly indicates the alkaline nature of the water. The dissolved oxygen, which is an important chemical factor that reflects the health and productive status of a water body, ranged from 7.20 to 10.30 mg/l, being maximum in December and minimum in July. The slight increased in dissolved oxygen contents during winter months may be due to overall cumulative effects of low temperature and increase in transparency. The availability of free carbon dioxide ranged from 0.80 to 1.45 mg/l *i.e.* negligible, which is another indicator of suitability of water for fish production. The presence of free carbon dioxide in negligible quantity that too during summer shows poor rate of decomposition at the bottom. The carbonate values were observed to be poor ranged from zero in summer to maximum of 12.0 mg/l in December, whereas, the bicarbonate values ranged from 99.4 mg/l to 216.50 mg/l being minimum in January and maximum in September.

The value of chloride contents were observed between 0.98 to 5.98 mg/l. As the chloride contents were very low in the reservoir reflects that there is very less amount of organic waste of animal origin. The calcium, magnesium and sulphate were observed with in the range from 14.30 to 22.80 mg/l; 2.10 to 9.96 mg/l and 11.62 to 34.56 mg/l respectively. On the basis of above observation, it may be assumed that the reservoir water is free from any kind of pollution.

The period of one year investigations revealed the occurrence of 13 species of phytoplanktons and 9 species of zooplanktons. The phytoplankton recorded from the reservoir belongs to chlorophyceae, Bacillariophyceae and Cynophyceae. Chlorophyceae was emerged as largest group and was represented by 7 genera viz; *Volvox, Microspora, Uronema, Spyrogyra, Pediastrum, Closteridium* and *Treubaria.* Bacillarophyceae is represented by *Pinnularia, Navicula, Gomphonema, Synedra* and *Tabellaria.* Whereas the cynophyceae was represented by one genera i.e, *Oscillatoria* species. The phytoplankton population was found to increase during post monsoon and reached to maximum during winter. Kant and Kachroo (1979) have also recorded winter peak in phytoplankton population while studying the Mansar lake. The decrease in phytoplankton population during the summer and Monsoon as observed in salal reservoir may be ascribed to drifting impact of rain during monsoon along with high turbidity and dilution of water as earlier reported by Lund (1965). Similar observation were also made by Pehwa and Mehrotra (1966), Nautiyal (1985) and Shyam Sunder (1983), while studying the different aspects of river Ganga, river Alaknand and river Jehlum respectively.

Zooplankton of the reservoir was represented by 9 genera, belonging to two groups viz; rotifera and cladocera. The rotifers were represented by *Brachious, Keratella, Anuraeopsis, Lepedella, Monostyla,Testidunella* and *Lecane* species; where as the cladocerans were represented by *Macrothrix* and *Allona.* The presently observed increased pattern of rotifers during post monsoon period is in concurrence with the observation of Govind (1969), who have observed increased pattern of rotifer population during post monsoon and attributed it to the reproduction and multiplication of member species during this period. The member of cladocera recorded their peak during winter and minima during monsoon. The fall in numerical count of zooplankton during monsoon may be due to the drifting of these organisms under the impact of accelerated water currents resulted as a consequence of water influx from the catchment areas. Turbidity caused by surface run off and soil erosion due to rains too severely effect the zooplankton population as earlier reported by Joshi (1996).

In total 10 species of fishes *viz., Tor putitora, Tor tor, Schizothorax plagiostomus, S. progastus, Glyptothorax pectinopterus, Nemacheilus* spp, *Crossocheilus* spp, *Garra gotyla, Garra lamta* and *Barilius bendelisis* have been reported from the reservoir. The paucity of fish fauna in the reservoir may be due to mountain barrier separating the area from rest of the state. The fish yield from the reservoir is also low, which can be ascribed due to irrational use of fishing gears and large scale exploitation of brooders and juveniles in the past.

Table 16.3: List of Fishes Reported from Salal Reservoir

1.	*Tor putitora*	2.	*Tor tor*
3.	*Schizothorax plagiostomus*	4.	*Schizothorax progastus*
5.	*Glyptothorax pectinopterus*	6.	*Namaichelus* sps.
7.	*Crossocheilus* sps.	8.	*Garra gotyla gotyla*
9.	*Garra lamta*	10.	*Barilius benedilius*

It has been observed that among the existing fish fauna, the mahseer fishery in the reservoir is vanishing due to construction of dam, which affects the migratory behaviour of the *Tor* Species, since there is no fish pass. The construction of dam has created a complete barrier for fish to migrate up stream. Earlier this fish use to ascend nullah Rud and Ans for breeding purpose and therefore suffered to a severe set back due to construction of dam. Earlier, the mahseer was found in abundance in the area, but now, the catches reveals that the mahseer catch is declining. However the previous authentic data on fish catch is not available, but the personal observation and local inquiries revealed that the fish population of the Mahseer is dwindling gradually and is at the verge of extinction, if certain conservative steps are not taken. This fish can certainly be protected by adopting suitable conservation and rehabilitation measures and if need felt, Mahseer seed is produced and raised upto the size of fingerlings can be released in the reservoir under river matching programme, so as to conserve the *Tor putitora*. For this purpose, the State Fisheries Department has well established mahseer hatchery at Anji (Reasi), where the artificial promotion of *Tor putitora* is undertaken by stripping method. In addition to propagation of mahseer, the strict conservation measures involve the check of illegal fishing of brooders and juveniles. Observance of close season if declared would facilitate further propagation of the fish. The propagation of mahseer seed on large scale and stocking of the same in the other selected areas will go a long way to increase the mahseer population.

Considering the urgent requirement to boost fish production from this reservoir, major thrust should be on stocking on a well designed approach. Stocking of quality fish seed on scientific line is another important tool in reservoir fisheries management. When the reservoir is formed, new trophic niches are created and indigenous fish faunas will not be able to utilize these niches and hence suitable species needs to be stocked. Such species should be introduced after thorough investigation of behaviour, adaptability, growth and reproduction, which can acclimatize in prevailing biotic and abiotic factors. The species should be non-competitive to indigenous species but should have fast growth rate. The stocking of reservoir with proper number and size of yearlings of cultivable indigenous fishes like rohu,catla,mrigal is the essential feature for the development of the reservoir fisheries. The stocking of yearlings of proper size and also with the desirable fishes to some extent would solve the problems of survival, so that the stocking stock became a part of natural fishery for ensuring the continuity of their kind. The yield from the reservoir can also be enhanced by adopting pen and cage culture. The fish species suitable for stocking are *Catla catla, Labeo rohita; L.calbasu; Cirrhina* sps. However, the exotic carps, common as well as Chinese *i.e. Cyprinus carpio* sps, (Common carp) *Ctenopharyngodon idella* (Grass carp) and *Hypophthalmichthys molitrix* (Silver carp) have not found suitable for introduction as the introduction of the common carp may disturb the indigenous fauna; whereas the chinese carp have raised a dust of controversy in case of Govindsagar reservoir (Krishna Rao, 2005).However, it is advisable to follow integrated fisheries, as it will be a feasible proportion to combine the activity with poultry, dairy, duckery and horticulture.

To conclude, important measures such as strict vigilance to check capture of brooders and juveniles, closed season fishes and Judicious exploitation through use of desirable net are some of the important measures to ensure effective conservation and management of the reservoir for enhancing the fish yield. The management should be based on fish stock manipulation, though it is advisable to restore the selective stocking for correcting the imbalance, if any, in species composition. Regular stocking of the reservoir with fast growing, high yielding and adaptable species of yearlings for improving the density will definitely step up fish production. However, the precautions should be taken to select and stock such species which could cover all the niches of the reservoir and also to reorient the quality and quantity of the fish stock for better fish yield.

Recommendations

The reservoir being virgin since no exotic species has been added so far. It is recommended that the reservoir be exclusive developed for sport fisheries, where mahseer could be the main thrust area. There are hardly few water bodies left, where sport loving Angler's could get "Thrilling Mahseer" particularly " Golden Mahseer" *Tor putitora* to play with. This reservoir which is still a virgin reservoir with beautiful Golden Mahseer; if managed properly by adopting suitable conservation measures and management methods could become a beautiful spot for Angler's by way of putting sincere efforts. The mahseer which has got trapped in this man made water body and if able to acclimatize/ establish in this new environment, and also in two beautiful streams; which were earlier also providing very good breeding grounds, could not only preserve the endangered mahseer but also develop the recreational fishery and enhance the fish yield. Besides, *Schizothorax sps*, another important fish for angling as well as the commercial value could also be established.

References

APHA, 1995. *Standard Methods for Examination of Water and Wastewater*. APHA, New York, Washington.

Battish, S.K., 1992. *Freshwater Zooplankton of India*. Oxford and IBH Publishing Co. Pvt. Ltd., New Delhi.

Golterman, H.L. and Clymo, 1971. *Methods for Chemical Analysis of Freshwaters*. Blackwell Scientific Publications, Oxford.

Govind, B.V., 1969. Bottom fauna and macro vegetation in the Tungabhadra reservoir and their role in the food chain of fish communities. In: *Seminar on the Ecology of Freshwater Reservoir*, Sponsored by ICAR and CIFRI, Barrakpore.

Jayaram, K.C., 1998. *Freshwater Fishes of Indian Region*. NPH, New Delhi.

Kant, S. and Anand, V.K., 1979. Interrelationship of Phytoplankton and physical factors in Mansar Lake, Jammu (J and K). *India Ecol.*, 5(2): 134–140.

Krishna Rao, D.S., 2005. Strategies for enhancement of fish production from Indian reservoirs. *Fishing Chimes*, 25(1): 86–90.

Lund, J.W.G., 1963. The ecology of Fresh water Phytoplankton. *Bio. Rew.*, 40: 231–239.

Joshi, C.B., 1996. Hydrobiological studies on the river Kallagi in Keralla. *Indian J. Fish*, 23(1 and 2): 72–73.

Jyoti, M.K., Sharma, K.K., Baba, D.I. and Mediha, S., 2004. Fishery potential of India reservoirs with special reference to Ranjit Sagar Dam. In: *Proceedings of National Symposium on Status of Cold Water Fisheries with Reference to Fragile Aquatic Ecosystem*. 15th All India Congress of Zoology 29th–31th October, P.G. Deptt. Of Zoology, University of Jammu.

Malhotra, Y.R., Jyoti, M.K. and Dutta, S.P.S., 1975. An aid to the identification of fishes found in Jammu division of Jammu and Kashmir State. *Jammu University Review*, 5: 50–56.

Nautiyal, P.1985. Studies on the riverine ecology of torrential waters in India uplands of the Garhwal region–III, Florestic and Faunistic survey. *Trop. Ecol*, 27:157–165.

Needham, J. G. and Needham, P.R., 1962. *Freshwater Biology*. Holden Day Inc. Son. Francicco.

Pehwa, D.V. and Mehrotra S.N., 1968. Observations on fluctuation of in abundance of planktons in relation to hydrobiological conditions of river Ganga. *Proc. Nat. Acad. Sci. India*, 53(B) 2:157–189.

Pennak, R.W., 1978. *Freshwater Invertebrates of United States*. John Wiley and Sons, New York, 803 pp.

Sakhare, V.B. and Joshi, P.K., 2004. Present status of reservoir Fisheries in Maharashtra. *Fishing Chimes*, 24(8): 56–60.

Sunder, S., 1983. Studies on phytoplankton and other algae of India, Sagar tank, Udaipur, India. *Hydrobiol.*, 31: 421–434.

Sugunan, V.V. and Sinha, M., 2000. *Guidelines for Small Reservoir Fisheries Management*. Bull No. 93 CIFRI, Barrackpore (WB).

Talwar, P.K. and Jhingran, A., 1991. *Inland Fishes of India and Adjacent Countries*. Oxford and IBH Publishing Co. Pvt. Ltd. New Delhi.

Chapter 17

Effect of Slaughter House Waste Based Diets on Growth, Feed Utilization and Body Carcass Composition of *Labeo rohita* (Hamilton) Fingerlings

Pradeep Kumar Singh and Sandhya Rani Gaur

Introduction

Aquaculture is a feed based industry with over 60 per cent of the operational cost coming from feed sources alone (Pandian *et al.*, 2001). Fish meal is used as the major source of animal protein in most of the commercial fish and crustacean feed. However, due to the escalating cost and uncertain availability of reliable quality of fish meal, it is essential that alternative sources of protein be identified for its partial or complete replacement (Gallanger,1994). With the increasing research thrust on this aspect during recent years there has been much interest on the possibility of using various non-conventional feed sources. Next to fishery by-products, terrestrial vertebrate by-product usually constitutes the second major source of animal protein within aquafeeds for warm water fish species (Tacon, 1993a). The utilization of most vertebrate by–product meal is usually limited by specific nutritional imbalances including imbalances of essential amino acids and ash mineral (Tacon, 1993b). Another advantage of using these animal by-products is their easy availability in the locality and low cost. Several workers have attempted to replace fishmeal in diet with slaughter house waste (Nandeesha *et al.*, 1991; Jadhav and Rao, 1991; Khan and Jafri, 1992 and Hassan *et al.*, 1999) for Indian Major Carps.

In the present investigation, an attempt has been made to evaluate the efficacy of slaughter house waste for their possible inclusion in the diet of *Labeo rohita*.

Materials and Methods

Experimental Site

The present experiment was conducted at Aquaculture lab, Department of Fisheries, Raipur and feed analysis part was done at Department of Animal Nutrition, Anjora, Durg of Indira Gandhi Agricultural University, Raipur, Chhattisgarh, India during period of December to February, 2003-04.

Preparation of Experimental Diets

Five different nearly iso-caloric (3.6-3.7 kcal/g) diets with 23.96, 25, 30, 35 and 40 per cent crude protein were formulated. D_1 to D_4 were prepared by finely powdered, mustard oil cake, rice bran, slaughter house waste, vitamin- mineral mixture and soybean oil with carboxy methyl cellulose as a binder, while control diet was prepared by mixing all the above except slaughter house waste. The proximate composition of the ingredients used in formulated of the diets is given in Table 17.1. The formulation and proximate composition of the control and the test diets are displayed in Table 17.2.

Table 17.1: Proximate Composition of Feed Ingredients Used in Formulation of Different Diets (g/100g, as fed basis)

Ingredients	*Moisture*	*Crude Protein*	*Crude Fat*	*Crude Fibre*	*Total Ash*	*NFE*
Rice bran	9.0±0.16	13.56±0.36	10.22±0.11	19.33±0.19	10.53±0.07	37.36
Mustard oil cake	8.5±0.10	38.00±0.50	8.44±0.12	7.5±0.10	14.06±0.04	23.50
Slaughter house waste	7.8±0.20	65.00±1.88	6.11±0.25	2.33±0.17	7.93±0.02	10.83

Table 17.2: Ingredients (%) and Proximate Composition of the Experimental Diets (% on Dry Matter Basis)

Ingredients ($g100g^{-1}$)	*Diets*				
	Control	D_1	D_2	D_3	D_4
Rice bran	46.5	58.38	48.68	38.96	29.25
Mustard oil cake	46.5	20.00	20.00	20.00	20.00
Slaughter house waste	–	14.62	24.32	34.04	43.75
Soybean oil	2.0	2.0	2.0	2.0	2.0
Vitamin-Mineral mixture*	2.0		2.0	2.0	2.0
Carboxy methyl cellulose	3.0	3.0	3.0	3.0	3.0
Proximate composition **					
Moisture	7.50	8.30	8.20	8.15	8.35
Crude protein	23.97	25.00	30.00	35.00	40.00
Crude fat	10.67	10.55	10.15	9.74	9.16
Crude fibre	12.46	13.12	11.47	9.82	8.17
Total ash	11.42	10.10	9.95	9.61	9.36
Nitrogen free extract		32.93	30.23	27.68	24.96
Gross Energy (kcal $100g^{-1}$)	361.08	363.34	367.22	369.13	369.93

* Vitaminate forte, Roche India Ltd.; ** On dry matter basis.

Experimental Design

The *Labeo rohita* fingerlings were obtained from the Fish Farm, Department of Fisheries IGAU, Raipur and maintained in glass aquaria (60×30×30 cms) in aquaculture lab. Fingerlings were acclimatized for a period of 15 days prior to the commencement of experiment. The aquaria water was renewed weekly. During the acclimation period the fingerlings were fed on control diet.

Fingerlings (mean body weight 1.83±0.3g) were randomly distributed @ 10 fish per aquarium with four replicates of each dietary treatment. All fish were fed daily twice a day at 08.30- 09.00 h and 1600-1630 h, the feeding being 4 per cent BW/day. The fish were bulk weighed every week and feeding rate adjusted accordingly. Fish were exposed to the respective diet for 3h during each ration; therefore, the uneaten feed was siphoned out, stored and dried separately for calculating the feed conversion ratio. At the termination of the experiment, the fish from all the treatments were weighed individually and processed for subsequent analysis.

Analytical Methods

Water quality was monitored at weekly interval for temperature, pH, dissolved oxygen, free carbon-dioxide and total alkalinity (APHA, 1989).

Prior to formulation of diets, all the feed ingredients were analysed for their proximate composition (Table 17.2) using standard methods (AOAC, 1990). Difference in weight after keeping the sample in oven (100°C) for 24 hrs. indicating the moisture content. Ash content was determined by incinering the sample in muffle furnace (600°C) for 5 hrs. Crude protein by micro- kjeldhal method (N X 6.25), while fat was quantified by continuous Soxhlet extraction technique using petroleum as solvent; Crude fibre by fibra plus apparatus.

Data Collection

Absolute growth = Final weight – Initial weight

Relative growth rate (RGR) (per cent/d) = Absolute growth/initial wt. × 100

Specific growth rate (SGR) = Log_e Final wt.–Log_e Initial wt./Experimental period (days) × 100

Food conversion ratio (FCR) = Total dry feed fed/total weight gain

Protein efficiency ratio (PER) = Wet weight gain of fish (g)/Protein consumed

Protein productive value (PPV) = Net increase in carcass protein/Protein consumed ×100

Survival rate per cent = Total number of surviving fish/Total number of fish stocked ×100

Statistical Analysis

Analysis of variance (Snedecor and Cochran, 1968) followed by Duncan's Multiple Range Test (Duncan, 1955) was employed to test the differences between the means.

Results

The water quality parameters (temperature, pH, dissolved oxygen, total alkalinity and free CO_2 of the water in the experimental aquaria ranged from 18.7-26.2°C, 5.6-6.8, 6.65-7.20 mg L^{-1}, 80-100 mg L^{-1} and 0.0-2.32 mg L^{-1} respectively) were well within the range for warm water fish culture.

Growth performance indicated by absolute growth, relative growth rate and specific growth rate (SGR) when subjected to analysis of variance revealed significant influence of slaughter house waste (Figure 17.1 and Table 17.3). Data indicate that the fish fed with diet D_2 having 30 per cent protein level

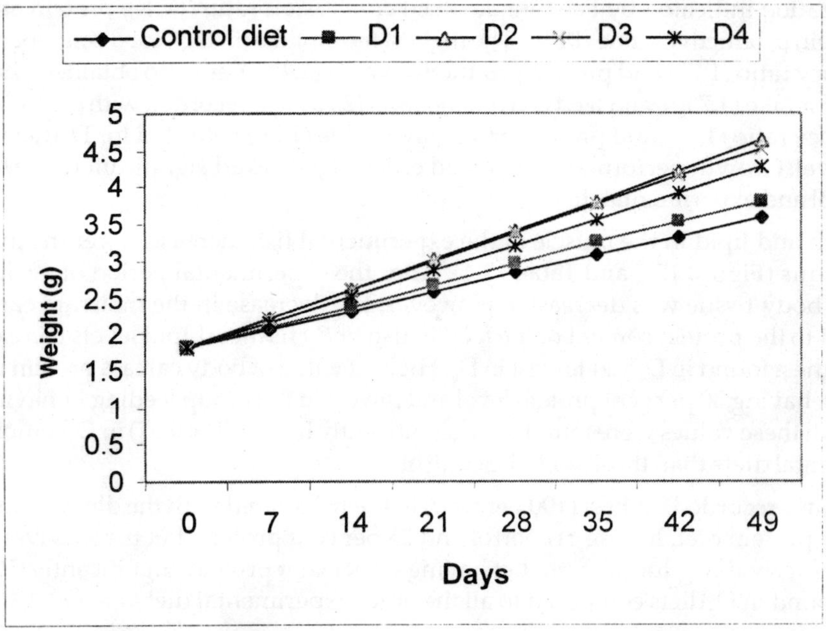

Figure 17.1: Weekly Growth of Rohu Fingerlings Under Different Dietary Treatments

reflected best growth in terms of absolute growth (2.80g), relative growth rate (153.27) and specific growth rate (2.48). The lowest absolute growth (1.75), relative growth rate (96.15) and specific growth rate (1.55) were recorded with control diet.

Table 17.3: Growth and Feed Utilization of *Labeo rohita* Fingerlings Fed the Control and Test Diets for 49 Days

Parameters	*Diets*				
	Control	D_1	D_2	D_3	D_4
Initial length (mm)	55±12.5	55±14.2	55±13.4	55±14.5	55±15.2
Initial weight (g)	1.82±0.01	1.82±0.01	1.83±0.02	1.81±0.01	1.81±0.01
Final length (mm)	69±18.4	71±19.2	74±25.1	72±22.2	71±21.4
Final weight (g)	3.58±0.01[e]	3.79±0.01[d]	4.63±0.02[a]	4.51±0.01[b]	4.27±0.02[c]
Absolute growth (g)	1.75±0.01[e]	1.96±0.01[d]	2.80±0.01[a]	2.69±0.01[b]	2.44±0.02[c]
Relative growth rate	96.15±0.02[e]	107.39±0.01[d]	153.27±0.01[a]	148.52±0.02[b]	133.52±0.01[c]
Specific growth rate	1.55±0.01[e]	1.73±0.01[d]	2.48±0.01[a]	2.38±0.02[b]	2.16±0.07[c]
Feed intake	4.79±0.01[e]	4.92±0.02[d]	5.32±0.01[a]	5.21±0.01[b]	5.07±0.01[c]
Feed conversion ratio	2.73±0.04[a]	2.50±0.02[b]	1.89±0.01[d]	1.93±0.01[d]	2.07±0.01[c]
Protein efficiency ratio	1.53±0.01[c]	1.61±0.03[b]	1.75±0.07[a]	1.48±0.06[c]	1.19±0.11[d]
Protein Productive Value	2.06±0.02[c]	2.28±0.03[b]	2.48±0.07[a]	1.61±0.04[c]	1.29±0.06[d]
Survival (per cent)	97.5±2.5[a]	97.25±2.5[a]	100±0.0[a]	100±0.0[a]	95.0±2.88[a]

±SEM; Values with the same superscript in same rows are not significantly different (P<0.01).

Feed utilization indicated by Feed intake, feed conversion ratio (FCR), protein efficiency ratio (PER) and protein productive value (PPV). The highest feed intake (5.32g), food conversion ratio (1.89), protein efficiency ratio (1.75) and protein productive value (2.48) were also obtained with the D_2 diet. The lowest feed intake (4.79g), and feed conversion ratio (2.73) was recorded with control diet, whereas protein efficiency ratio (1.19) and protein productive value (1.29) recorded for D_4 diet having 40 per cent protein level. Growth performance and feed efficiency showed significant ($P < 0.01$) difference between control and experimental diets.

The protein and lipid in the muscle of the experimental fish increased over the initial in all the dietary treatments (Figure 17.2 and Table 17.4). After the experimental period of 49 days, moisture content in fish body tissue was decreased. However, the decrease in the moisture content was not directly related to the protein content of the feed in fish body tissues. Lipid levels increases in all diet groups and highest found in D_1 and lowest in D_4. Highest value of body carcass protein was observed in fish fed on D_2 having 30 per cent protein level and lowest in the group feeding D_4 having 40 per cent protein level. All these values were found to be significantly higher ($P < 0.01$) in D_2 compared to all the other experimental diets than those with the control diet.

Survival rates recorded highest (100 per cent) in the fish reared with the diet D_2 and D_3 having 30 and 35 per cent protein diet. In case of control and 25 per cent protein diet the survival rate was 97.5 per cent. Least survival was found with diet having 40 per cent protein. Significantly ($P<0.01$) higher survival was found in D_2 diets compared to all the other experimental diets.

Table 17.4: Carcass Composition of Fish Before and After the Feeding Experimental Diets During Experimental Period

Nutrient	*Initial Value*	*Diets*				
		Control Diet	*D_1*	*D_2*	*D_3*	*D_4*
Moisture (%)	83.42±0.23	78.75±0.21^{bc}	78.24±0.23^{c}	78.64±0.28^{bc}	79.16±0.23^{a}	78.94±0.23^{ab}
Dry matter (%)	27.58±0.34	21.25±0.21^{bc}	21.76±0.23^{a}	21.36±0.34^{b}	20.84±0.23^{bc}	21.06±0.41^{bc}
Crude protein (% DM)	55.71±0.13	58.04±0.08^{c}	58.34±0.11^{b}	59.21±0.25^{a}	58.34±0.10^{b}	57.17±0.20^{d}
Crude fat (% DM)	10.90±0.09	13.17±0.07^{b}	13.53±0.08^{a}	13.21±0.05^{b}	12.99±0.12^{c}	12.52±0.17^{d}
Total ash (% DM)	20.17±1.1	15.61±0.03^{ab}	15.57±0.06^{bc}	15.65±0.05^{a}	15.58±0.04^{bc}	15.55±0.04^{bc}

±SEM; Values in the same rows with different superscripts differ significantly ($P < 0.01$).

Discussion

At the end of the feeding trial, significant increase in weight gain was recorded among the various treatment groups fed diets containing slaughter house waste as compared to control diet. This indicates the efficacy of slaughter house waste as a feed ingredient for *Labeo rohita*. The highest percentage of live weight gain indicating best performance was noted with fish fed on D_2 having 30 per cent protein level in diet. The finding point to the fact that under a given condition better performance in terms of growth, digestion, absorption and eventually better conversion of nutrients to flesh. Greater degree of protein (Jafri and Anwar, 1995) and energy (Jafri and Hassan, 1998) digestibility of goat offal by *Labeo rohita* justifies the performance of this diet in producing maximum live weight gain. In a review, Nandeesha *et al.* (1991) stated that the growth attained by carps on slaughter house waste based diet was comparable to that fed on fish meal. Hassan *et al.* (1999) also found best growth of *Labeo rohita*

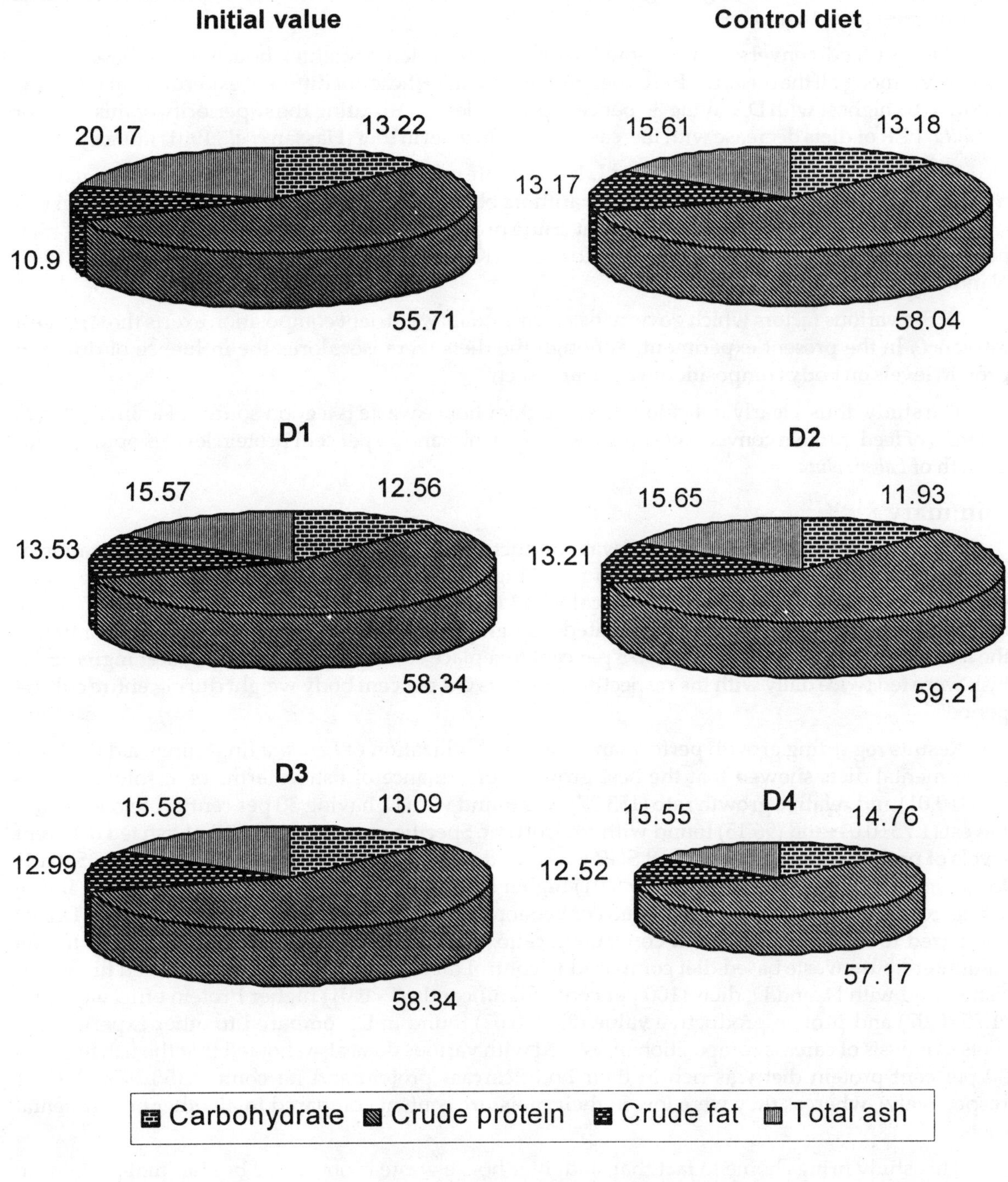

Figure 17.2: Initial and Final Carcass Composition (% Dry Matter Basis) of *Labeo rohita* Under Various Diet Treatments

fingerlings with diet containing slaughter house waste. Jadhav and Rao (1991) reported that goat offal gives the best growth in carps.

The best feed conversion was noted with the group fed slaughter house waste based diets. Similarly, among all the diets, the PER, used to compare the efficacy of different experimental feed, was found to be highest with D_2 having 30 per cent protein level, indicating the superiority of this feed for *L. rohita*. PER of diets decrease with increase of protein level in diet (Hassan *et al.*, 1991; Jauncey 1982).

Results on optimum protein levels for *Labeo rohita* are in broad agreement with those of (Gangadhara *et al.*, 1997; Chakraborty *et al.*, 1999). These authors obtained high growth and feed efficiency in fish when fed on a diet containing 30-35 per cent crude protein. Further, growth depressing effect of high protein levels in carps has been reported for rohu (Khan and Jafri, 1992; Chakraborthy *et al.*, 1994; Nandeesha *et al.*, 2000).

Of the various factors which govern the carcass quality, nutrient composition exerts the strongest influence. In the present experiment, although the diets were isocaloric, the influence of different protein levels on body composition was clearly seen.

The study, thus, clearly indicated that slaughter house waste is a good source of animal protein in terms of feed, protein conversion and growth efficiency and 30 per cent protein level is optimum for growth of *Labeo rohita*.

Summary

A seven week feeding experiment was conducted to evaluate the suitability of incorporation of slaughter house waste in the diet of rohu fish. Fingerlings of *Labeo rohita* (weight 1.83±0.02g) were used as experimental fish. Five nearly isocaloric (3.6-3.7 kcal/g) diets with 23.96 (control), 25, 30, 35, and 40 per cent protein level were formulated. Slaughter house waste in the diets was incorporated at the levels of 14.62, 24.32, 34.04 and 43.75 per cent to replace equal proportion of the other ingredients. Fish were fed twice daily with the respective test diets @ 4 per cent body weight during entire culture period.

Results regarding growth performance and feed utilization of *L. rohita* fingerlings fed different experimental diets showed that the best growth performance of fish in terms of absolute growth (2.80±0.01) and relative growth rate (153.27) was found with D_2 having 30 per cent protein level and lowest (1.75±0.01) and (96.15) found with control diet. Specific growth rate (SGR) of fish fed different levels of protein showed that highest SGR was found with D_2 diet (2.48±0.01) and lowest (1.55±0.01) found in control diet. Significantly ($P<0.01$) higher SGR was found in diets having slaughter house waste, compared to control diet. The feed conversion ratio (1.89±0.01) was lowest in fish fed on D_2 diet compared to all other diets. Feed conversion ratio was significantly ($P<0.01$) lower in fish fed on slaughter house waste based diet compared to control diet. Survival rates were recorded highest in fish reared with D_2 and D_3 diets (100 per cent). Significantly ($P<0.01$) higher Protein efficiency ratio (1.75±0.07) and protein productive value (2.48±0.07) found in D_2 compared to other experimental diets. Analysis of carcass composition of fish fed with various diets also showed that the fish fed with 30 per cent protein diet was rich in their body carcass protein and fat content (58.34 and 13.21 respectively), whereas they were low in their moisture content compared to all other experimental diets.

This study brings home to fact that slaughter house waste is one of the best animal proteins in terms of growth, food conversion and body carcass composition and 30 per cent protein level is optimum for growth of *Labeo rohita*.

Acknowledgements

The authors thank Dr. S.P. Tiwari, Professor and Head, Department of Animal Nutrition, College of Veterinary Sciences and Animal Husbandry and Dr. Shailendra Shrivastava, Incharge Fish Farm, College of Agriculture, Indira Gandhi Agricultural University Raipur, Chhattisgarh, India for their encouragement and facilities provided.

References

AOAC, 1990. *Official Methods of Analysis*, 15th Edition, (Ed.) K. Helrich. Association of Official Analytical Chemists, Arlington VA, 684 p.

APHA, 1989. *Standard Methods for the Examination of Water and Wastewater*, 17th Edition. American Public Health Association, New York, 1527 p.

Chakraborty, S.C., Chowdhury, S.A. and Chakraborty, S., 1999. Energy budget of Indian Major Carp, *Labeo rohita* fingerlings fed on diets with different protein levels. *Asian Fish. Science*, 12: 297–308.

Duncan, G.B.,1955. Multiple range and multiple F-test. *Biometrics*, 11: 1–42.

Gangadhare, B., Nandeesha, M.C., Varghese, J.J. and Keshavanath, P.,1997. Effect of varying protein and lipid levels on the growth of rohu, *Labeo rohita*. *Asian Fisheries Science*, 10: 139–147.

Hassan, M.A., Jafri, A.K. and Khan, M.A.,1999. Efficacy of animal wastes as supplementary feed for Indian Major Carp. *Labeo rohita*, fingerling. *J. Inland Fish. Soc., India*, 31(2): 52–57.

Jadhav, A.G. and Prasad, Rao, P.D., 1991. Effect of supplementary feed such as abattoir and fish market waste on growth of the fingerlings of *Labeo rohita* (Ham.) In: *Proc. National Sem. on New Horizons in Freshwater Aqua*. Central Institute of Freshwater Aquaculture, Bhubaneswar, 23–25 Jan. pp. 81–83.

Jafri, A.K. and M.A. Hassan, 1998. Energy digestibility coefficiemt of commonly used feedstuffs in different size class of Indian Major Carps, *Labeo rohita* (Hamilton) and *Cirrhinus mrigala* (Hamilton). *Asian Fish. Science*, 12(2): 153–163.

Jafri, A.K. and M.F. Anwar, 1995. Protein digestibility of some low cost feedstuffs in fingerlings of Indian Major Carps. *Asian Fish. Science*, 8: 47–53.

Jauncey, K., 1982. The effect of varying dietary protein levels on the growth, food conversion, protein utilization and body composition of juveniles tilapias (*Sarotherodon mossambicus*). *Aquaculture*, 27: 43–54.

Khan, M.A. and Jafri, A.K., 1992.Replacement of fish meal with offal meal in formulated diets for the carp fingerlings *Labeo rohita* (Ham.). *J. Aquaculture*, 2: 56–62.

Nandeesha, M.C., Gangadhara, B., Varghese, T.J. and Keshavanath, P., 1991. Growth response and flesh quality of common carp, *Cyprinus carpio* fed with high levels of nondefatted silkworm pupae. *Asian Fisheries Science*, 13: 235–242.

Pandian, T.J., Mohanty, S.N. and Ayyappan, S., 2001. Food requirements of fish and feed production in India. In: *Sustainable Indian Fisheries*, (Ed.) Pandian, T.J. National Academy of Agriculture Science, New Delhi, pp. 153–165.

Rangacharyulu, P.V., Paul, B.N., Nandi, S., Sarkar, S. and Mukhopadhyay, P.K., 2000. Effect of different protein and energy levels on growth, nitrogen metabolism and body composition of rohu (*Labeo rohita*). *J. Aqua.*, 8: 17–24.

Snedecor, G. W. and Cochran, W.G., 1968. *Statistical Methods*. Oxford and IBH Publishing Company, New Delhi, India, 593p

Tacon, A.D.J., 1993. Feed formulation and on-farm feed management, pp. 61–74. In: *Proceedings of the FAO/AADCP*. Regional Expert consultation on farm made aquafeeds, (Eds.) M.B. New A.D.J. Tacon and I. Savas, 14–18 December, 1992, Bangkok, Thailand. FAO/AADCP, Bangkok, Thailand, 434 pp.

Tacon, A.D.J., 1993b. *Feed Ingredients for Warm Water Fish: Fish Meal and other Processed Feedstuffs. FAO Fisheries Circular* No. 856, FAO, 64 p.

Chapter 18

Quantitative Protein Requirement of the Fingerlings of Indian Major Carp Rohu, *Labeo rohita* (Hamilton)

Pradeep Kumar Singh, Sandhya Rani Gaur and M.S. Chari

Introduction

The Indian Major Carps are highly nutritious and most prestigious of all fishes in India. Among the Indian Major Carps, rohu (*Labeo rohita*) is most preferred species and constitute about 35 per cent of the Indian Major Carps production (FAO, 2000). Aquaculture is a feed based industry with over 60 per cent of the operational cost coming from feed sources alone (Pandian *et al.*, 2001). The cost of feed is largely influenced by the level and sources of protein which is the most expensive component of a fish diet. It is the major dietary components which influence growth of fish. Insufficient as well as excess level of protein in feed is not desirable, the former results in poor growth, while the latter would be wasted by diverting for energy. Hence, dietary protein level in fish feed needs to be optimized accurately. However, studies on the dietary protein requirements of carps feeding on dry formulated feeds are few (Renukardhya and Varghese, 1986; Mohanty *et al.*, 1990; Seenappa and Devaraj, 1995; Chakraborty *et al.*, 1999; Rangacharyulu *et al.*, 2000). In the present study, the optimum dietary protein requirements for optimum growth, food utilization and body carcass composition of *Labeo rohita* feeding on four slaughter house waste based diets having 25 per cent, 30 per cent, 35 per cent, and 40 per cent protein content have been investigated.

Materials and Methods

The present experiment was conducted at Aquaculture Lab, Department of Fisheries, Raipur and feed analysis part was done at Department of Animal Nutrition, Anjora, Durg of Indira Gandhi Agricultural University, Raipur, Chhattisgarh, India during period of December to February, 2003-04.

Four experimental diets with 25, 30, 35, and 40 per cent protein level respectively were prepared and named as D_1, D_2, D_3 and D_4 respectively. The ingredients selected were slaughter house waste, mustard oil cake, rice bran, soybean oil, vitamin-mineral mixture and carboxy methyl cellulose. For feed preparation, the ingredients were grounded separately in an electric grinder and sieved to remove large particles. The required quantity of feed ingredients boiled with water in a pressure cooker for 30 minutes. The boiled mixture was allowed to cool and then vitamin–mineral mixture was added and mixed well. These mixtures are processed through a hand pelletizer for preparing pellets, which were then dried in room temperature for two days. The proximate composition of the ingredients used for formulation of diets is given in Table 18.1.The composition of ingredients used for experimental diets given in Table 18. 2. The proximate compositions of the experimental diets are displayed in Table 18.3.

Table 18.1: Proximate Composition of Feed Ingredients Used in the Formulation of Diets

Ingredients	*Moisture*	*Crude Protein*	*Crude Fat*	*Crude Fibre*	*Total Ash*	*NFE*
Rice bran	9.0±0.16	13.56±0.36	10.22±0.11	19.33±0.19	10.53±0.07	37.36
Mustard oil cake	8.5±0.10	38.00±0.50	8.44±0.12	7.5±0.10	14.06±0.04	23.50
Slaughter house waste	7.8±0.20	65.00±1.88	6.11±0.25	2.33±0.17	7.93±0.02	14. 58

±SEM.

Table 18.2: Composition of Experimental Diets

Ingredients	*Diets*			
	D_1	D_2	D_3	D_4
Rice bran	58.38	48.68	38.96	29.25
Mustard oil cake	20.00	20.00	20.00	20.00
Slaughter house waste	14.62	24.32	34.04	43.75
Soybean oil	2.00	2.00	2.00	2.00
Carboxy methyl cellulose	3.00	3.00	3.00	3.00
Vitamin-Mineral mixture*	2.00	2.00	2.00	2.00

Table 18.3: Proximate Composition of the Experimental Diets (per cent on dry matter basis)

Parameters	*Diets*			
	D_1	D_2	D_3	D_4
Moisture	8.30	8.20	8.15	8.35
Crude protein	25.00	30.00	35.00	40.00
Crude fat	10.55	10.15	9.74	9.16
Crude fibre	13.12	11.47	9.82	8.17
Total ash	10.10	9.95	9.61	9.36
Metabolizable energy kcal/100g^{-1}	363.34	367.22	369.13	369.93

Composition of Vitamin–mineral Mixture

Vitamin A (as acetate) 5000 I.U.; Cholecalciferol 1000 I.U.; Thiamin monocitrate 10.00mg; Riboflavin 10.00 mg; Pyridoxine hydrochloride 5.00 mg; Cyanacobalamin 15.00 g; Nicotinamide 75.00 mg; Calcium pentathenate 10.00 mg; Ascorbic acid 150.00 mg; a tocopheryl acetate 25.00 mg; Biotin 5.00 mg; Folic acid 5.00 mg; Menadione 100.00 mg ; Choline chloride 50.00 mg; PABA 5.00 mg; Myoinocitol 10.00 mg; Calcium lactate 0.125 mg; Magnesium oxide 60.00 mg; Dried ferrous sulphate 30.00 mg; Copper sulphate 2.00 mg; Manganese sulphate 2.00 mg; Zinc sulphate 2.00 mg; Sodium molybdate dihydrate 0.25 mg; Sodium borate 0.80 mg; Potassium iodate 20.00 mg; Bicalcium phosphate 0.10 g; Cobalt chloride 20.00mg.

The acclimatised *Labeo rohita* fingerlings (mean body weight 1.83±0.02g) were randomly distributed @ 10 fish per aquarium with four replicates of each dietary treatment. All fish were fed daily twice a day at 08.30- 09.00 h and 16.00-16.30 h, the feeding being at 4 per cent body weight day^{-1}. The aquaria water was partially renewed daily. The fish were weighed at weekly intervals and feeding rate adjusted accordingly. Fish were exposed to the respective diet for 3h during each ration and the uneaten feed was siphoned out, stored and dried separately for calculating the food conversion ratio (FCR).

Water quality was monitored at weekly interval for temperature, pH, dissolved oxygen, free carbon di–oxide and total alkalinity (APHA, 1989). Feeding ingredients, experimental diets and the fish muscle were analysed for the proximate composition following standard methods (AOAC, 1990).

Data Collection

Weight gain (g) = Individual Final weight–Individual Initial weight/Individual Initial weight × 100

Specific growth rate (SGR) = Log_e final weight – Log_e initial weight/Experimental duration (days) × 100

Food conversion ratio (FCR) = Feed intake (g)/Live weight gain (g)

Protein efficiency ratio (PER) = Wet weight gain of fish (g)/Protein consumed

Survival rate = Total number of surviving fish/Total number of fish stocked ×100

Data were tested for significance employing one–way analysis of variance (Snedecor and Cochran, 1968) and Duncans (1955) multiple range test.

Results

Physico-chemical Parameters of Water

The water quality parameters in different treatments were within the optimal range for carp. Temperature ranged from 19.7 to 27°C, with dissolved oxygen fluctuating between 5.63 to 7.41 mg/l^{-1}. Water pH ranged from 7.45- 8.82, total alkalinity from 80 to 100 mg/l^{-1} and free carbon di-oxide fluctuating between 0.0-2.32mg/l^{-1}.

Growth Performance

Growth performance indicated by absolute growth, percentage increase in live weight and specific growth rate (SGR) and survival rate when subjected to analysis of variance revealed significant influence of protein level (Table 18.5). Data indicate that the fish fed with diet having 30 per cent protein (D_2) reflected best growth in terms of live weight gain (153.2 per cent) with the specific growth rate being

2.48 per cent. The lowest live weight gain (96.26) and specific growth rate (1.55 per cent) were recorded with D_1 having 25 per cent protein. Growth performance and feed utilization showed significant ($P < 0.01$) difference between D_2 and D_1. Similar trends were also found with D_2 and D_4 and between D_2 and D_3.

Table 18.4: Weekly Growth Rate of *Labeo rohita* Fingerlings Under Different Protein Levels

Diets	*Weight (g/fish) (in weeks)*							
	0	*1*	*2*	*3*	*4*	*5*	*6*	*7*
D_1	1.82±0.01	2.11±0.02	2.38±0.01	2.67±0.02	2.95±0.01	3.25±0.02	3.52±0.01	3.79±0.01
D_2	1.83±0.02	2.22±0.01	2.61±0.02	3.01±0.01	3.41±0.02	3.81±0.01	4.22±0.01	4.63±0.02
D_3	1.81±0.01	2.21±0.02	2.60±0.01	2.99±0.02	3.37±0.01	3.76±0.01	4.15±0.02	4.51±0.01
D_4	1.82±0.01	2.17±0.01	2.51±0.02	2.86±0.02	320±0.01	3.56±0.02	3.91±0.01	4.27±0.02

±SEM.

Weekly weight gains of rohu fingerlings fed with different diets are presented in Table 18.4. The average initial and final weight of rohu fingerlings were between 1.81 to 1.83g and 3.58 to 4.63g respectively in different dietary treatment. Highest growth was found with D_2 having 30 per cent protein and lowest with D_1 having 25 per cent protein level.

Food Conversion Ratio and Protein Efficiency Ratio

The food conversion ratio (FCR) and protein efficiency ratio (PER) are also influenced by the protein level in the diets (Figure 18.1). The food consumption and wet weight production plays an important role in the increase or decrease of food conversion ratio. In the present experiment the diet with 30 per cent protein level showed the minimum FCR whereas maximum FCR was recorded for the rohu fed with 25 per cent protein diet.

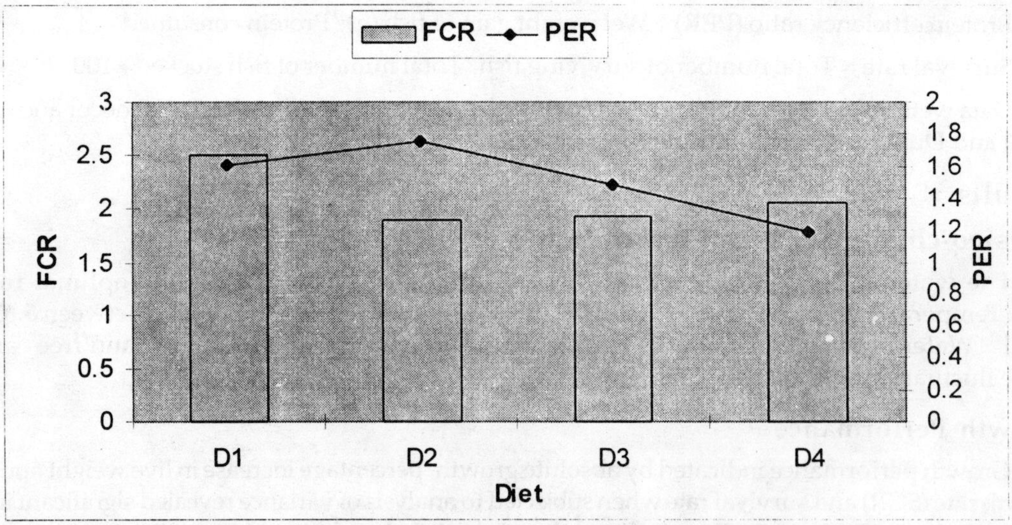

Figure 18.1: Food Conversion Ratio and Protein Efficiency Ratio of *Labeo rohita* Fed on Diets having Different Protein Levels

The highest PER at 30 per cent protein level in the diet indicates that protein is efficiently utilized and growth was also better than all the other diet groups. The low PER values were recorded above 30 per cent protein level.

Carcass Composition

The protein and lipid in the muscle of the experimental fish increased over the initial in all the dietary treatments (Table 18.5). The fish fed with 30 per cent protein level resulted highest amount of protein and lipid in the muscle.

Table18.5: Effect of Protein Levels on the Growth and Muscle Composition of *Labeo rohita* Fingerlings

Details	*Diets*			
	D_1	D_2	D_3	D_4
Initial weight	1.82±0.01	1.83±0.02	1.81±0.01	1.81±0.01
Final weight	3.79±0.01	4.63±0.02	4.51±0.01	4.27±0.02
Net Weight gain	1.96±0.01 ^{d}	2.80±0.01 ^{a}	2.69±0.01 ^{b}	2.44±0.02 ^{c}
Weight gain	107.39±0.01 ^{d}	153.27±0.02^{a}	148.52±0.02 ^{b}	133.52±0.01 ^{c}
Specific Growth Rate	1.73±0.01^{d}	2.48±0.01^{a}	2.38±0.02^{b}	2.16±0.01^{c}
Food Conversion Ratio	2.50±0.02 ^{a}	1.89±0.01 ^{d}	1.93±0.01 ^{c}	2.07±0.01 ^{b}
Protein Efficiency Ratio	1.61±0.03 ^{b}	1.75±0.07 ^{a}	1.48±0.06 ^{c}	1.19±0.11
Survival rate (per cent)	97.25±2.5 ^{a}	100±0.0 ^{a}	100±0.0 ^{a}	95.0±2.88 ^{a}
Muscle composition				
Protein (Initial value: 55.71)	58. 34±0.11^{b}	59. 21±0.25^{a}	58. 34±0.10^{b}	57.17±0.20^{c}
Lipid (Initial value: 10.90)	13. 21±0.15^{b}	13. 53±0.08^{a}	13.17±0.18^{b}	12. 52±0.17^{c}

±SEM; Values in the same rows with different superscripts differ significantly ($P < 0.01$) .

Survival

Survival rates were recorded highest in the reared with 30 per cent and 35 per cent protein diet. Survival rate decreased with increased or decreased level of protein.

Discussion

Present studies on Indian Major Carp, *Labeo rohita* has revealed high growth, and high accumulation of carcass protein in fish fed on a diet containing 30 per cent protein.

The present study shows that specific growth rate and food conversion ratio in *Labeo rohita* are influenced by the dietary protein content. Studies on effect of protein level on growth performance of fishes clearly indicate that while feeding with slaughter house waste based diets gives optimum growth occurs between 30-35 per cent dietary protein levels. The present results recorded the maximum growth performance by *Labeo rohita*, feeding on 30 per cent dietary protein and hence lie within the range reported by other workers for rohu and other fishes (Renukardhya and Varghese, 1986; Seenappa and Devaraj, 1995; Gangadhara *et al.*, 1997; Chakraborty *et al.*, 1999; Rangacharyulu *et al.*, 2000 and Paul and Mohanty, 2002).

The finding point to the fact that under a given condition better performance in growth means, better digestion, better absorption and eventually better conversion of nutrients to flesh. Greater degree of protein (Jafri and Anwar, 1995) and energy (Jafri and Hassan, 1999) digestibility of goat offal has been reported for *Labeo rohita* justify the performance of this diet in producing maximum live weight gain.

FCR and PER are known to be decreased with increasing dietary protein content (Jauncey, 1982) similar trends were found in present experiment.

Of the various factors which govern the carcass quality nutrient composition protein and lipid level exerts the strongest influence. In the present experiment, although the diets were isocaloric and different protein level, the influence of different protein levels on body composition was clearly seen. In conclusion, it appears that 30 per cent dietary protein level gave best result in terms of growth, feed conversion, survival and protein utilization and improving the flesh quality of rohu fingerlings.

In conclusion, it appears that 30 per cent dietary protein level is optimum for attaining maximum growth, feed conversion, survival and protein utilization and improving the flesh quality in *Labeo rohita* fingerlings.

Conclusion

A forty-nine day feeding experiment conduced to determine protein requirement of fingerlings of *Labeo rohita* (average weight 1.83±0.02g) feeding on four formulated diets with varying protein levels (25, 30, 35 and 40 per cent) using slaughter house waste as the major protein source. In terms of growth, food conversion ratio, protein efficiency ratio, survival and ratios of protein and lipid deposition in muscle, diet containing 30 per cent protein level revealed a significantly ($P<0.01$) better performance of the fish in comparison with the fish fed on other diets containing lower or higher protein levels. These findings suggest that about 30 per cent crude protein with 370 kcal/100g energy content in diet appears to be sufficient for obtaining optimum growth in *Labeo rohita*.

Acknowledgements

The authors are indebted to Dr. S.P. Tiwari, Head, Department of Animal Nutrition, I.G.A.U., Anjora (Durg) and Dr. Shailendra Shrivastava, Farm Incharge, Department of Fisheries, I.G.A.U., Raipur for extending laboratory facilities.

References

AOAC, 1990. *Official Methods of Analysis*, 15th Edition, (Ed.) K. Helrich. Association of Official Analytical Chemists, Arlington VA, 684 p.

APHA, 1989. *Standard Methods for the Examination of Water and Wastewater*, 17th Edition, American Public Health Association, New York, 1527 p.

Chakraborty, S.C., Chowdhury, S.A. and Chakraborty, S., 1999. Energy budget of Indian Major Carp, *Labeo rohita* fingerlings fed on diets with different protein levels. *Asian Fisheries Science*, 12: 297–308.

Duncan, G.B., 1955. Multiple range and multiple F-test. *Biometrics*, 11: 1–42.

FAO, 2000. *Fishery Statistics (Aquaculture Production)*. Food and Agriculture Organization of the United Nations, Rome, 90(2): 22–26, 36, 38, 114, 131.

Gangadhare, B., Nandeesha, M.C., Varghese, J.J. and Keshavanath, P., 1997. Effect of varying protein and lipid levels on the growth of rohu, *Labeo rohita*. In: *Asian Fisheries Science*, 10: 139–147.

Jafri, A.K. and M.A. Hassan, 1999. Energy digestibility coefficiemt of commonly used feedstuffs in different size class of Indian Major Carps, *Labeo rohita* (Hamilton) and *Cirrhinus mrigala* (Hamilton). *Asian Fish. Science,* 12(2): 153–163.

Jafri, A.K. and M.F. Anwar, 1995. Protein digestibility of some low cost feedstuffs in fingerlings of Indian major carps. *Asian Fish. Science,* 8: 47–53.

Jauncey, K., 1982. The effect of varying dietary protein levels on the growth, food conversion, protein utilization and body composition of juveniles tilapias (*Sarotherodon mossambicus*). *Aquaculture,* 27: 43–54.

Khan, M.A. and Jafri, A.K., 1992. Replacement of fish meal with offal meal in formulated diets for the carp fingerlings *Labeo rohita* (Ham.). *J.Aquaculture,* 2: 56–62.

Mohanty, S.N., Swamy, D.N. and Tripathy, S.D., 1990. Protein utilization in Indian major carp fry, *Catla catla* (Ham.) and *Labeo rohita* (Ham.) fed four protein diets. *J. Aqua. Trop.,* 5: 173–179.

Pandian, T.J., Mohanty, S.N. and Ayyappan, S., 2001. Food requirements of fish and feed production in India. In: *Sustainable Indian Fisheries,* (Ed.) Pandian, T.J. National Academy of Agriculture Science, New Delhi, pp. 153–165.

Paul, B.N., Mohanty, S.N., 2002. Recent advances in carp feeding in India. In: *IVth Biennial Conference and Exhibition,* Kolkata, pp. 42–48.

Rangacharyulu, P.V., Paul, B.N., Nandi, S., Sarkar, S. and Mukhopadhyay, P.K., 2000. Effect of different protein and energy levels on growth, nitrogen metabolism and body composition of rohu (*Labeo rohita*). *J. Aqua.,* 8: 17–24.

Renukardhya, K.M. and Varghese, T.J., 1986. Protein requirement of carps, *Catla catla* (Ham.) and *Labeo rohita* (Ham). In: *Proc. Indian Acad. Science (Anim. Sci.),* 95(1): 103–107.

Seenappa, D. and Devaraj, K.V., 1995. Effect of different levels of protein, fat and carbohydrate on growth, feed utilization and body carcass composition of fingerlings in *Catla catla* (Ham.). *Aquaculture,* 129: 243–249.

Snedecor, G.W. and Cochran, W.G., 1968. *Statistical Methods.* Oxford and IBH Publishing Company, New Delhi, India, 593 p.

Chapter 19

Comparative Study of Effect of Three Different Types of Food on Population Density of Freshwater Rotifer *Monostyla* sp.

Sampada Ketkar and Madhuri Pejaver

Introduction

Rotifers are valuable live food for larval fish and crustacean culture. In the study related to culture of dwarf gouramis, a species known for its micro-fry, Lim and Wong (1997) demonstrated that during first food feeding, rotifer fed fry grew up to 33 per cent faster and in addition they have a 44 per cent greater survival rate compared to yolk fed cohorts. Rotifers are also used as live feed for red and black sea bream (internet) as well as Southern flounder (Cool *et al.*, 2002). With the remarkable developments in larval rearing technology of industrial useful fishes, demand for the rotifer is further increasing. Practical applications have proved that the produced rotifers are suitable for larvicultures of marine shrimp, crab and fishes. The most important requirement to succeed in rotifer cultures is to provide the animals with a suitable food source (Pourriot, 1980). The dominant marine algal species are good food sources for rotifer growth. The algae can provide rotifers with sufficient protein and especially essential fatty acids. The bacterial populations in these ponds can provide rotifers with proteins and especially essential vitamins such as VB_{12} for rotifers growth (Yu *et al.*, 1988). But at the same time some species of rotifers appeared to be highly selective in their feeding habits (Pourriot, 1980), hence during present study, among bacteria and microalgae, which is more preferable as a food for mass culture was studied. For present study *Monostyla sp.* were tried to culture under laboratory conditions by using three techniques dependent on types of food. These techniques are as follows:

1. By using microalgae, Chlorella as food
2. By using bacteria from supernatant liquid of deoiled groundnut cake (DOGC) dipped in distilled water as food.
3. By using bacteria from supernatant liquid of rice bran dipped in distilled water as food.

Methodology

To study the effect of these food items on desired species of rotifers, experiments were arranged in three parallel sets. Each set consisted of five sterile glass beakers of 10 ml each. For three sets of experiments three different food items were used. Food used in each set is as follows:

1st set: *Chlorella*

2nd set: Bacteria in supernatant liquid of DOGC dipped in distilled water

3rd set: Bacteria in supernatant liquid of rice barn dipped in distilled water

To start the experiment, 5ml distilled water was taken in each beaker. Desired species of rotifer was inoculated as one organism per ml *i.e.* total 5 organisms in the beginning. Quantity of food fed to organisms in each set of experiments was constant *i.e.* 0.1 ml per beaker. Each set was repeated five times. Each set was observed at the interval of 6 hrs and total no. of organisms were recorded by sub sampling method till the end of the experiment ie. after the death of all organisms.

For counting total number of organisms sub sampling method was used. 1 ml sub sample was taken in separate beaker and was observed under microscope. Density was counted and the procedure was repeated three times. After counting organisms were returned back to the original culture.

At the end of experiments average readings were calculated and are represented in respective tables

The experiments where 'sudden deaths' was observed, were repeated with little modification to justify the 'death'. These repeated experiments were started as mentioned earlier but after every eight days subcultures were made with density of two individuals per ml *i.e.* total ten organisms in the beginning and continued till same result was obtained.

Results and Discussion

During the present study, the maxima of population density of *Monostyla sp.* was obtained in different days in different sets of experiments (Table 19. 1). The maximum of population density of first set with the *Chlorella* as food was observed at the end of 28 days. In second set with bacteria in supernatant liquid of DOGC dipped in the distilled water as food it was observed at the end of 31 days while in third set it was observed at the end of 29 days where bacteria in supernatant liquid of rice barn dipped in the distilled water was used as food.

Table 19.1: Population Density of *Monostyla sp* in Different Media (Food)

Day	No. of Organisms/5ml		
	Chlorella	*DOGC*	*Rice Bran*
0	5	5	5
1	7	18	13
2	10	30	25
3	18	48	35
4	28	62	42
5	32	78	55
6	43	90	65
7	52	109	78
8	56	125	85
9	64	143	93
10	72	165	105
11	81	185	115
12	90	207	125
13	98	225	134
14	106	245	145
15	113	267	156
16	122	290	165
17	133	312	178
18	139	332	185
19	145	355	197
20	153	370	209
21	162	395	217
22	168	415	225
23	178	435	235
24	186	450	247
25	195	468	256
26	203	490	262
27	203	508	262
28	203	520	262
29	All dead	520	262
30	–	520	All dead
31	–	520	–
32	–	All dead	–

Among these experiment sets, maximum population density *i.e.* 104 organisms/ml (520 organisms/5ml) was observed in the second set where bacteria in supernatant liquid of DOGC dipped in distilled water were used as food, while minimum population density *i.e.* 41 organisms/ml (203 organisms/5ml) was observed in first set with the *Chlorella* as the food. In third set intermediate population density *i.e.* 52 organisms/ml (262 organisms/5ml) was observed where bacteria in the supernatant liquid of rice barn dipped in distilled water as food.

According to Figure 19.1, it was observed that there was gradual increase in population density for first few days. Then steady population density before death of all organisms for last 2 to 4 days and then sudden death of all organisms in all sets of experiments was recorded.

As sudden death was observed in every set of experiment after gradual increase in the population density, to justify the death subcultures were made after every 8 days with each type of food. Subcultures were made after every 8 days *i.e.* 8th, 16th and 24th day with inoculation density two organisms per ml. Subcultures with *Chlorella* as the food were named as SCC1, SCC2 and SCC3 respectively and parent culture was named as PCC. Density was observed after every day and recorded in the Table 19.2. While subcultures with bacteria in the supernatant liquid of DOGC dipped in the distilled water as the food were named as SCD1, SCD2 and SCD3 respectively and parent culture was named as PCD. Density was observed after every day and recorded in the table (Table 19.3) and subcultures with bacteria in the supernatant liquid of rice bran dipped in the distilled water as the food were named as SCR1, SCR2 and SCR3 respectively and parent culture was named as PCR. Density was observed after every day and recorded in the table (Table 19.4).

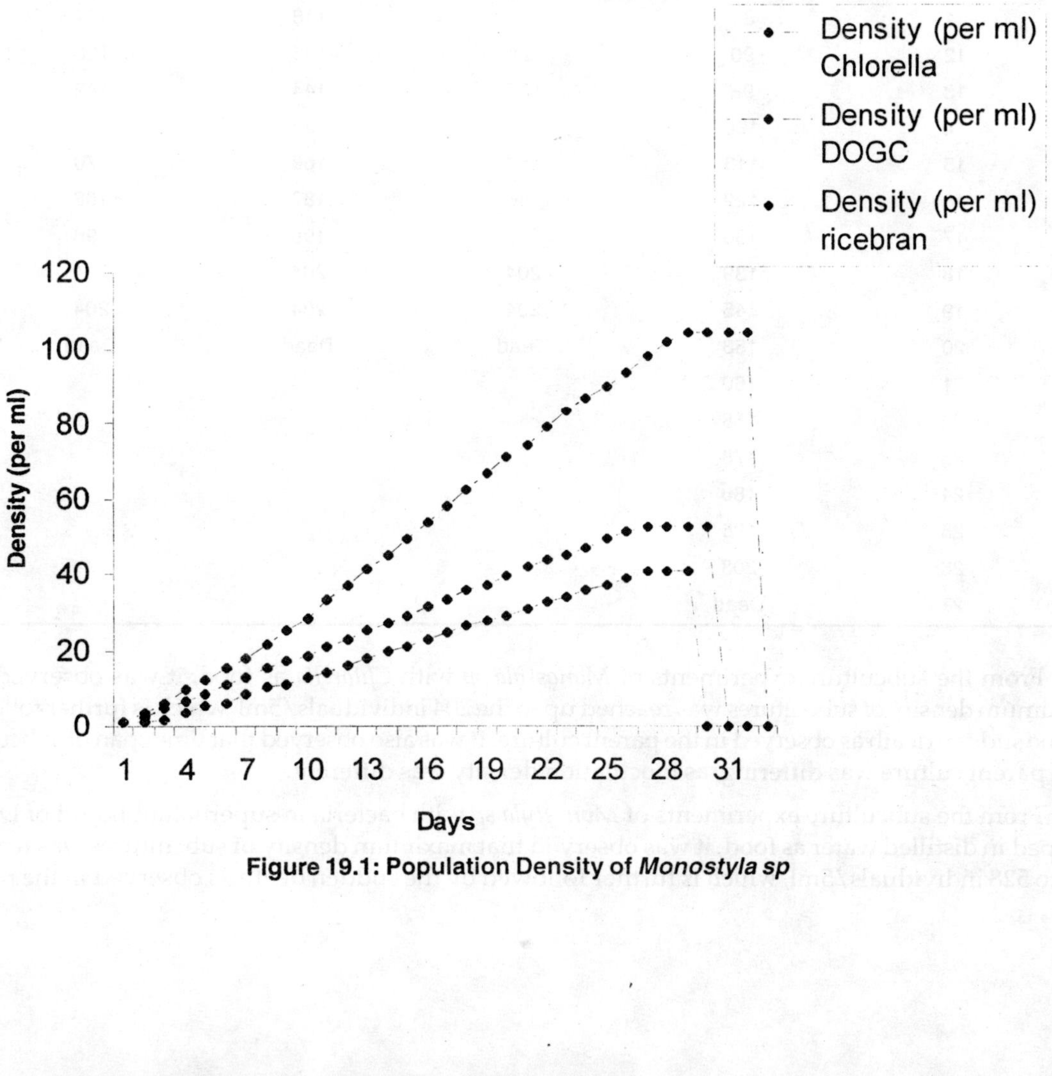

Figure 19.1: Population Density of *Monostyla sp*

Table 19.2: Comparative-density Count of *Monostyla* sp Subcultures and Parent Culture with *Chlorella* as Food

Day	*No. of Organisms/5ml*			
	PCC	*SCC1*	*SCC2*	*SCC3*
0	5	10	10	10
1	7	15	17	15
2	12	25	26	24
3	20	37	39	38
4	25	45	50	47
5	32	50	62	52
6	40	60	70	62
7	48	67	81	75
8	56	78	88	84
9	64	86	92	90
10	72	95	110	96
11	81	110	118	114
12	90	125	135	128
13	98	138	144	142
14	106	150	156	155
15	113	167	168	170
16	122	188	187	188
17	130	195	198	196
18	139	204	204	204
19	145	204	204	204
20	153	Dead	Dead	Dead
21	160			
22	168			
23	178			
24	186			
25	195			
26	203			
27	Dead			

From the subculture experiments of *Monostyla sp* with *Chlorella* as food, it was observed that maximum density of subcultures was reached up to the 204 individuals/5ml, which is further followed by the sudden death as observed in the parent culture. It was also observed that time span of subculture and parent culture was differing, as inoculation density was different.

From the subculture experiments of *Monostyla sp* with bacteria in supernatant liquid of DOGC dipped in distilled water as food, it was observed that maximum density of subcultures was reached up to 528 individuals/5ml, which is further followed by the sudden death as observed in the parent culture. It was also observed that time span of subculture and parent culture was differing, as inoculation density was different.

Table 19.3: Comparative Density Count of *Monostyla* sp Subcultures and Parent Culture with Bacteria In Supernatant Liquid of DOGC Dipped in Distilled Water as Food

Day	*No. of Organisms/5ml*			
	PCD	*SCD1*	*SCD2*	*SCD3*
0	5	10	10	10
1	20	42	44	42
2	30	60	62	59
3	50	98	100	101
4	65	134	135	135
5	80	178	178	180
6	94	200	205	205
7	109	220	222	220
8	125	255	254	252
9	143	278	280	281
10	165	305	304	304
11	185	335	338	340
12	207	395	398	398
13	225	420	425	428
14	245	468	470	470
15	267	490	495	494
16	290	520	525	526
17	312	528	526	528
18	332	528	526	528
19	355	528	526	528
20	375	528	526	528
21	395	Dead	Dead	Dead
22	415			
23	435			
24	455			
25	468			
26	488			
27	508			
28	528			
29	528			
30	528			
31	528			
32	All dead			

Table 19.4: Comparative Density Count of *Monostyla* sp. Subcultures and Parent Culture with Bacteria in Supernatant Liquid of Rice Barn Dipped in Distilled Water as Food

Day	*No. of Organisms/5ml*			
	PCR	*SCR1*	*SCR2*	*SCR3*
0	5	10	10	10
1	15	30	28	28
2	24	55	58	59
3	35	65	72	75
4	44	77	85	90
5	55	90	92	99
6	65	99	100	115
7	77	110	110	125
8	85	130	128	144
9	93	155	149	156
10	105	168	165	170
11	115	185	182	185
12	125	198	198	196
13	134	228	205	225
14	145	244	225	246
15	156	256	248	258
16	165	264	257	268
17	175	264	265	268
18	185	264	265	268
19	197	Dead	Dead	Dead
20	207			
21	217			
22	225			
23	235			
24	247			
25	256			
26	268			
27	268			
28	268			
29	268			
30	All dead			

From the subculture experiments of *Monostyla sp.* with bacteria in supernatant liquid of rice barn dipped in distilled water as food, it was observed that maximum density of subcultures was reached up to 268 individuals/5ml, which is further followed by the sudden death as observed in the parent culture. It was also observed that time span of subculture and parent culture was also differing, as inoculation density was different.

Hence these subculture experiments with each food on *Monostyla sp.* justified the reasons of sudden death.

The results during present study also show that as in the beginning when abundant food was available, there was steady growth of all organisms, for first few days. But at the end of experiments sudden death of all organisms was seen. This might be due to the following reasons.-

1. Unavailability of food
2. Accumulation of metabolic waste
3. Unavailability of sufficient amount of water
4. Overcrowding.

Yoshinga *et al.* (2003) reported that rotifers stopped active reproduction just after the onset of food deprivation and showed shorter subsequent survival times when they were starved at older ages. So this study can support the reasoning 'unavailability of food'. But at the same time the present study reveal that 'sudden death' can be a combined effect of all stated reasons.

From the present study, it can be concluded that for continuous culturing of these organisms, sub culturing will be effective which will also be useful for better yield. Rombant *et al.* (2003) demonstrated the use of dense nitrifying culture either in seed batch cultures, conventional bioreactors or hollow fiber bioreactor systems in support of rotifer cultures to be effective for improving the water quality and rotifer growth

The present investigations may also reveal that the sustainability of rotifer culture is dependant on the type of food used to feed the organisms and quantity of water available.Tadphale and Pejaver (2005) also reported the similar observations with respect to another genus *Lecane inopinata*.

Acknowledgements

The authors are thamkful to authorities of Live feed culture laboratories, CIFE, Mumbai for providing chlorella culture. The authors are also thankful; to staff and research scholars of department of zoololgy, B.N. Bandodkar college of science Thane.

References

Cool, R.W., Schwarz, M.H., Mowry, De and Jahncke, M.L., 2002. High density intensive rotifer culture for the first feed stage of juvenile Southern flounder (*Paralichthys lethostigma*). In: *Proc. 4th International Conference on Recirculating Co-operative Extension Service.*

Lim, L.C. and Wong, C.C., 1997. Use of the rotifer, Brachionus calyciflorus Pallas, in freshwater ornamental fish larviculture, *Hydrobiol*. Belgium: Kluwer Academic Publishers, 358: 269–273.

Pourriot, R., 1980. Workshop on culture techniques of rotifers. *Hyadrobiologia*, 73: 33–35.

Rombant, G., Grommen, R., Zizhong, Q., Vanhoof, V., Suantika, G., Dhert, P., Saorgeloos, P. and Verstraete, W., 2003. Include performance of an extensive rotifer culture system by using nitrifying inoculum (ABIL). *Aquacult. Res.*, 34(2): 165–174.

Tadphale, Sampada and Pejaver Madhuri, 2005. Preliminary observations on effect of three different types of food on the population density of the fresh water rotifer, Lecane inopinata, *J. Aquatic Biology*, 20(1): 159–161.

Yoshinga, T., Hagiwara A., and Tsukamoto, K., 2003. Life history response and age specific tolerance to starvation in *Brachionus plicatilis* O.F. Muller Rotifera. *J. Exp.Mar. Biol. Ecol.*, 287(2): 261–271.

Yu, J., A. Hino, R. Hirano and K. Hirayama, 1988. Vitamin B12-producing bacteria as a nutritive complement for a culture of the rotifer *Brachionus plicatilis. Nippon Suisan Gakkaishi*, 54: 1873–1880.

Chapter 20

Structure Immune Organ (Spleen) in *Mystus gulio* (Hamilton)

B. Deivasigamani

Introduction

The spleen in fishes is generally similar in structure and development to that of mammals. However, there are certain differences between mammalian spleen and fish spleen with regard to the structure and function. For example, capsular trabeculae system characteristic of mammalian spleen is not distinct in fish spleen (Zapata,1982). The white pulp in the spleen of most teleost is very limited and diffusely distributed. However, in some species distinct lymphoid accumulations are present around the arteries, veins and ellipsoids (Pitchappan,1980 and Pulsford *et al*, 1982).

Materials and Methods

Live specimens of *Mystus gulio* (Hamilton) of the family bagridae were collected from Ogyampet estuary near Chennai. Both sexes were used in the present study.The immune organ of spleen was dissected and removed to study their morphology and histology. The immune organ was fixed in Bouin's fluid for about 24 hours, processed, embedded in paraffin wax and cut into 6 µm thick sections. The sections were stained by Haematoxylin and eosin (HE) and Periodic acid-Schiff reaction (PAS).

Results and Discussion

The spleen of *Mystus gulio* is a reddish-brown, elongated, thick and flattened structure, lying along the intestine and in the proximity of the pancreas, trunk-kidney and the air-bladder (Figure 20.1). It measures about 5-10 mm in length and 3-5 mm in width, and weighs about 0.5-0.9 g. At one point on the surface of the spleen is a deep indentation, the hilum, which appears to divide the spleen into two halves (Figure 20.2). The splenic artery, a branch of the coeliacomesenteric artery, enters the organ through the hilum. Immediately, it divides into two, each traversing the respective half of the spleen. On entering into the splenic parenchyma, the arteries branch off repeatedly into a number of arterioles (Figure 20.3), that which ultimately end in capillaries. Several venules arising from different parts of the parenchyma merge into a larger vein that emerges out through the hilum (Figure 20.4), and joins the hepatic portal system. The veins and arteries run close to each other in most of the parenchyma (Figure 20.5). The spleen of *M. gulio* is covered by a very definite capsule of connective tissue (Figure 20.6). The capsule consists of closely packed fibrous tissue containing numerous elastic fibers and smooth muscles. The outer surface of the capsule appears either smooth (Figure 20.7), or somewhat serrated (Figure 20.8).

The spaces within the connective tissue framework are filled up with splenic pulp, the distinction of which into red pulp and white pulp is rather difficult in histological preparations. The pulp in general consists of large venous sinuses and their branches, representing the red pulp, and nodule-like ellipsoids of lymphoid tissue and dark-staining melanomacrophage centers, representing the white pulp (Figure 20.8).

Melanomacrophage centers (MMCs) in the spleen of *M. gulio* are aggregations of closely packed macrophages that contain yellow to dark-brown colored pigments (Figure 20.7). The size of these centers are highly variable ranging from very small (39 µm × 17 µm) consisting of a few cells to very large (131 µm × 103 µm) with hundreds of cells. The shape is mostly spheroid though elongated and bean shaped structures are not uncommon. The number of the MMCs is also variable, and it depends on the size *i.e.*, if they are larger they are fewer and if smaller, numerous (Figure 20.8).

The anatomy of the spleen of *Mystus gulio* is more or less similar to teleosts fishes. However, the presence of an hilum like indentation through which the artery and vein pass, in the spleen of *M. gulio* is significant in that such an observation has not been reported for any other teleosts species, and that it is a condition similar to that in mammals. The hilum divides the spleen into two lobes. In addition, insignificantly smaller lobes might also be encountered. The spleen of sharks and dogfish are much elongated and lobated. For instance, there are as many as 100 lobes in the spleen of the basking shark *Cetorhinus maximus* (Mathew and Parker,1950). The spleen of the spiny dogfish Squalus *acanthias* is less lobated (Fange and Nilsson (1985) and that of *Etmopterus spinax* consists of separate parts (Mattisson and Fange, 1982).

The morphology of the spleen in teleosts varies considerably. It may be a compact, unlobed and spindle shaped structure as in the cod *Gadus morhua* (Fange and Nilsson (1995) or much elongated and flat as in many species such as *T. mossabica* (Sailendri and Muthukkaruppan1975b), S. *gairdneri* (Grace and Manning,1980) and *D. labrax* (Quesada *et al*, 1990).

The weight of the spleen of *M. gulio* is about 0.5 to 0.9 g in fishes weighing 40 to 50 g, and it works out to be 0.01 to 0.02 per cent of the body weight. When compared to other species of teleosts, the spleen of *M. gulio* appears to be smaller. The weight of the spleen in percent of body weight is 0.05-0.14 in the bass *Lateolabrax japonicus* (Murata, 1959a), 0.12-0.34 in the mullet *Mugil cephalus* (Murata, 1959a), 0.04-0.86 in the Salmon *Salmo salar* (Miescher-Ruesch, 1881) and 0.02-0.18 in *G. morhua* and *Tinca tinca* (Fange and Nilson,1985). The spleen of holocephali and elasmobranchii are much larger and comparable to that of mammals.

A definite layer of connective tissue consisting of elastic fibers and smooth muscle cells encapsulates the spleen of *M. gulio*. Presence of a connective tissue capsule around the spleen has been reported in all the species studied and its nature is generally fibrous and elastic. The surface of the capsule in *M. gulio* appears somewhat serrated in many places, which may be foldings due to contraction of muscle

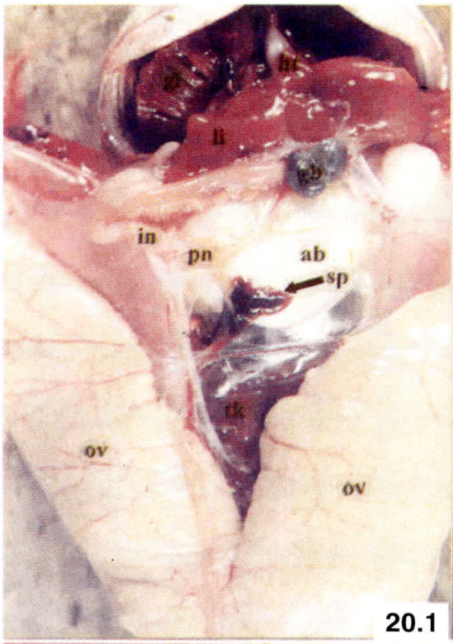

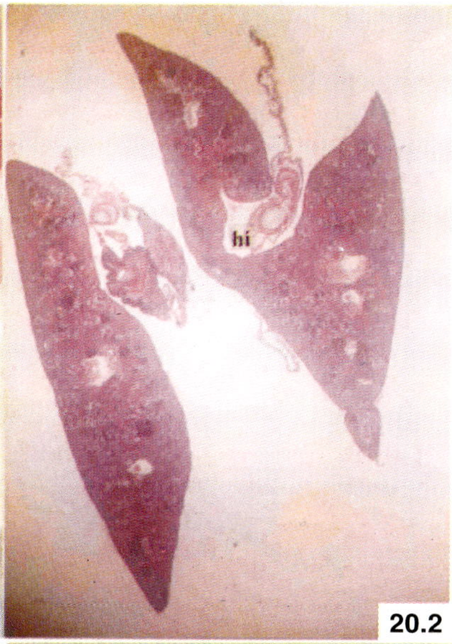

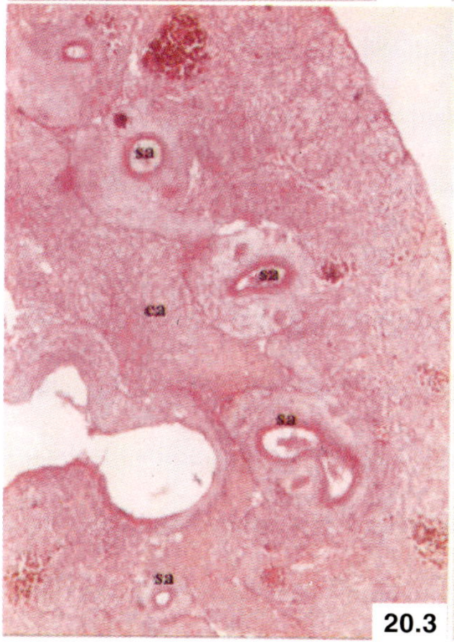

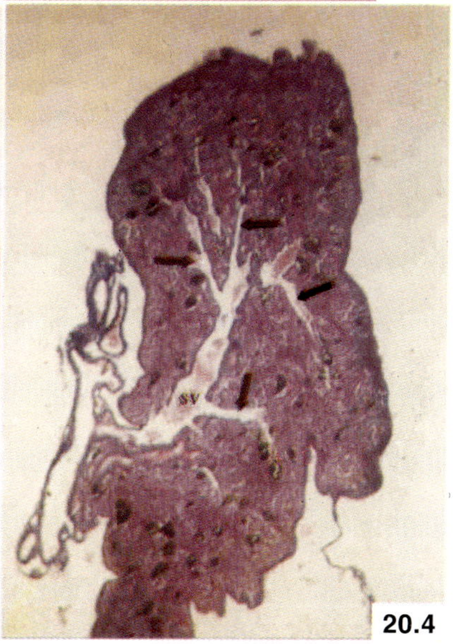

Figure 20.1: Dissection of *M. gulio* Displaying the Spleen in Relation to the Other Visceral Organs. × 1.025

Figure 20.2: Section of the Spleens from Two Specimens of *M. gulio* Showing in One of them the Entry of Splenic Artery through the Hilum that appears divide the organ into two halves. Bouin, HE. × 60

Figure 20.3: Showing the Splitting Up of the Splenic Artery into a number of Branches (Arrow) Bouin, PAS. × 600

Figure 20.4: Showing the Formation of Several Venules (Arrow) Joining the Splenic Vein that Emerges through the Hilum. Bouin, HE. × 60

ab: Air-bladder; ca: Capillary; gb: Gall bladder; gi: Gill; hi: Hilum; ht: Heart; in: Intestine; li: Liver; ov: Ovary, pn: Pancreas, sp: Spleen; sa: Splenic artery; sv: Splenic vein.

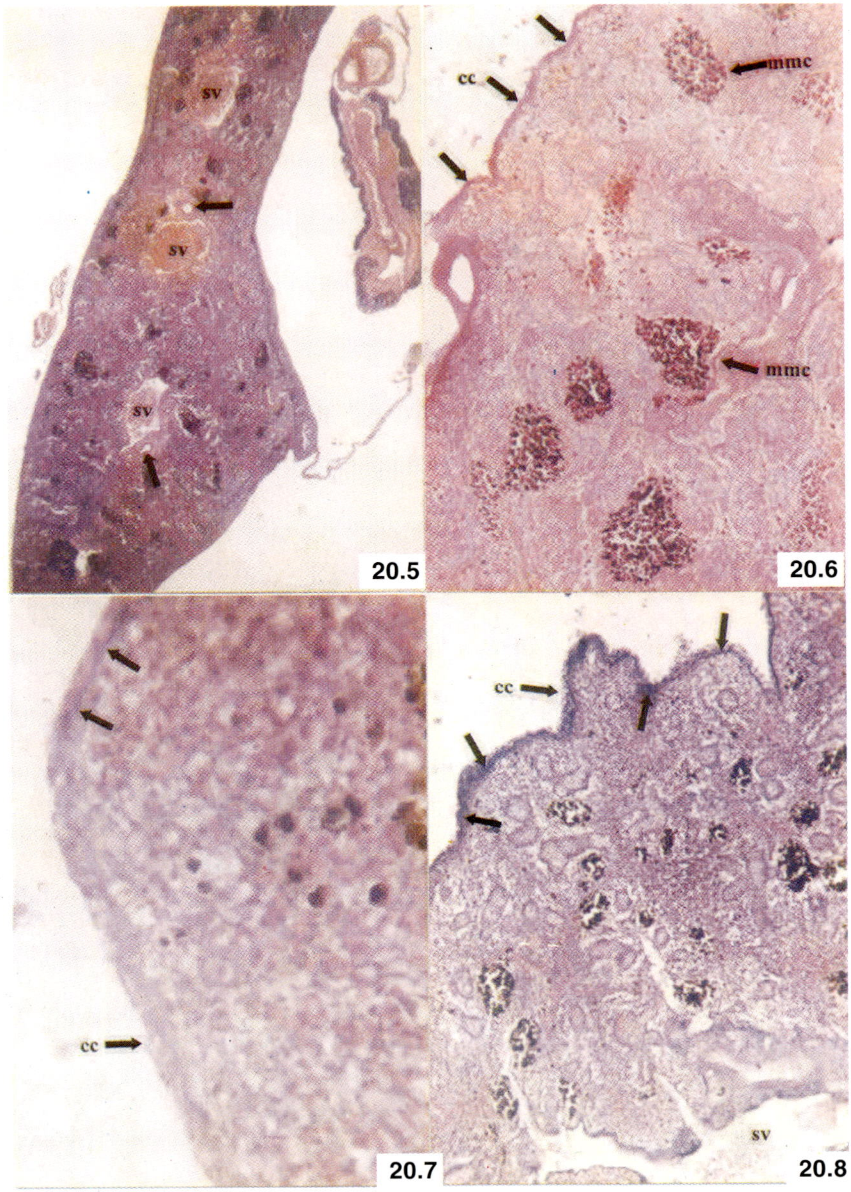

Figure 20. 5: Showing the Running of the Veins and Arteries (Arrow) Close to each other. Bouin, HE. × 60

Figure 20.6: Showing the Presence of a Definite Connective Tissue Capsule (Arrow) Consisting of Elastic Fibers. Bouin, PAS. × 130

Figure 20.7: Showing Even Condition of the External Surface of the Connective Tissue Capsule. Smooth muscle fibers (arrow) are seen on the inner lining of the capsule. Bouin, HE. × 600

Figure 20.8: Showing the Serrated Condition of the External Surface of the Connective Tissue Capsule (Arrow). Bouin, HE. × 130

cc: Connective tissue capsule; mmc: Melanomacrophage center; sv: Splenic vein.

fibers in the connective tissue. In mammals it is known that spleen is capable of changing its volume to store blood and expel it into circulation when required (Hartwig and Hartwig, 1985).

A well-developed trabecular system is characteristic of the spleen *M. gulio*. The connective tissue trabeculae consisting of the elastic fibers and smooth muscle cells extend from the capsule into the pulp, dividing the later into a number of incomplete anatomizing chambers. The spleens of other teleosts such as *T. mossambica* (Sailendri and Muthukkaruppan,1975b), *S. gairdneri* (Grace and Manning, 1980) and *C. carpio* (Secombes and Manning, 1980) also have trabecular system. The presence of a well-developed trabecular system consisting of elastic fibers and smooth muscle cells and the predominance of red pulp suggests that the spleen of *M. gulio* is also endowed with a function of storage and release of blood.

The splenic pulp of *M. gulio* includes a plexus of red pulp, and ellipsoids and MMCs representing the white pulp. The venous sinuses and the red pulp are erythrocytic in nature though a number of other cells such as monocytes, eosinophils etc. could be identified. Erythrocytes and thrombocytes are produced in the splenic red pulp of elasmobranchs (Fey, 1965 and Zapata, 1980). Granulopoiesis takes place in the spleen of the electric ray, *Torpedo ocellata* (Fey, 1965). In both elasmobranchs and teleosts lymphocytopoiesis occurs in the lymphoid areas of the spleen (Ellis *et al.*, 1976).

The spleen of *M. gulio* is the melano-macrophage centers that are nodules of pigment-containing macrophages. They are not membrane bound, and their size, shape and distribution are highly variable. They are mostly located close to ellipsoids often occupying the entire macrophage-lymphocyte area. While in some teleosts such as *Coryphaenoides rupestris* and *Argentina silus* the MMCs are well developed, in other species such as *Lophius piscatorius* and *Cyclopterus* these centers are indistinct or lacking (Fange and Nilson, 1985 and Herraez and Zapata, 1986). Sailendri and Muthukkaruppan (1975b) did not observe any MMCs in the spleen of *T. mossambica*.

Ellis *et al* (1976) describe the MMCs in the spleen of the plaice *P. platessa* as melanomacrophage aggregations surrounded by a cuff of white blood cells, mainly of lymphocyte type. The pigment of melanomacrophages is believed to consist of melanin, lipofuchsin and hemosiderin (Herraez and Zapata, 1986).

The observations in the present study on the histomorphology of the spleen of *M. gulio*, such as the well developed trabecular system, predominance of red pulp and close association between the ellipsoids, lymphoid tissue and MMCs full of pigmented macrophages, indicate that this organ might be involved in multiple functions such as storage and release of blood, phagocytosis, destruction of the blood cells and immune responses. However, as also obtained by Fange and Nilson (1985), the present knowledge on the structure and functions of the ellipsoids and MMCs, and the relationship between these structures and lymphoid tissue of the fish spleen is insufficient and therefore further investigations on a wide variety of species are needed to throw more light on these aspects.

Acknowledgements

The author is grateful to his research guide Dr. N. Gurumani, Head, Department of Microbiology, Pachaiyappa's College, Chennai–600 030, for providing the facilities to carry out the work and Dr. T. Balasubramanian, Director of CAS in Marine Biology, Annamalai University for his support and encouragement.

References

Ellis, A.E., A.L.S. Munroe and R.J. Roberts, 1976. Defence mechanisms in fish. *J. Fish Biol.*, 8: 67–78.

Fänge, R. and S. Nilsson, 1985. The fish spleen: structure and function. *Experientia*, 41: 152–158.

Fey, F., 1965. Haematologische Untersuchungen der blut-bildenden Gewebe niederer Wirbeltiere. *Folia Haemat.*, 84: 122–146.

Grace, M.F. and M.J. Manning, 1980. Histogenesis of the lymphoid organs in rainbow trout, *Salmo gairdneri*. *Dev. Comp. Immunol.*, 4: 255–264.

Hartwig, H., and H.G. Hartwig, 1985. Structural characteristics of the mammalian spleen indreating storage and release of red blood cells. Aspects of evolutionary and environmental demands. *Experientia*, 41: 150–163.

Herraez, M.P. and A.G. Zapata, 1986. Structure and function of the melano-macrophage centers of the gold fish, *Carassius auratus*. *Vet. Immunol. Immunopathol.*, 12: 117–126.

Mathews, L.H. and H.W. Parker, 1950. Note on the anatomy and biology of the basking shark, *Cetorhinus maximus*. *Proc. Zool. Soc. Lond.*, 120: 535–576.

Mattisson, A. and R. Fänge, 1982.The cellular structure of the Leydig organ in the shark, *Etmopterus spinax* L. *Biol. Bull.*, 162: 182–194.

Miescher-Ruesch, F., 1881. Uber das Leben des Rhein-Lachses in Susswasser. I. Abhandlung: Die Milz des Rhein-Lachses und ihre veranderungen. *Arch. Anat. Physiol.*, 193–218.

Murata, H., 1959a. Comparative studies of the spleen in submammalian vertebrates I. Topographical anatomy and relative weight of the spleen. *Okajimas Folia Anat. Jap.*, 33: 1–9.

Pitchappan, R.M., 1980. Review on the phylogeny of splenic structure and function. *Dev. Comp. Immunol.*, 4: 395–416.

Pulsford, A., R. Fänge, and W.J.W. Morrow, 1982. Cell types and interactions in the spleen of the dog fish, *Scyliorhinus canicula* L. *J. Fish Biol.*, 21: 649–662.

Quesada, J., M.I. Villena and B. Agulleiro, 1990. Light and electron microscopic study of the structure of the spleen of the sea bass (*Dicentrarchus labrax*). *J. Morphol.*, 206: 273–281.

Sailendri, K. and V.R. Muthukkaruppan, 1975b. Morphology of lymphoid organs in a Cichlid teleost, *Tilapia mossambica* (Peters). *J. Morphol.*, 147: 109–122.

Secombes, C.J. and M.J. Manning, 1980. Comparative studies on the immune system of fishes and amphibians. I. Antigen localization in the carp, *Cyprinus carpio*. *J. Fish Dis.*, 3: 399–412.

Zapata, A., 1980. Splenic erythropoiesis and thrombopoiesis in elasmobranches. An ultrastructural study. *Acta Zool., Stockh.*, 61: 59–64.

Zapata, A., 1982.Lymphoid organs of teleost fish. III. Splenic lymphoid tissue of *Rutilus rutilus* and *Gobio gobio*. *Dev. Comp. Immunol.*, 6: 87–94.

Chapter 21

Effects on Nucleic Acid Content of Muscles in the Starved Mosquito Fish *Gambusia affinis*

I.A. Raja and K.M. Kulkarni

Introduction

Fish of the family Poeciliidae, *Gambusia affinis* (Mosquito fish), were used because of their ease of handling and their resistance in conditions of culture and high importance in mosquito control (Fernando *et al.*, 1991), compared to other species of poecilidian fishes (Olivier and Kaiser, 1997).

They are resistant and energetic, and are quite active with smaller animals, (Mills, 1984; Mills and Vevers, 1986).Different workers studied effects of feeding and fasting on fish physiology and body contents (Garling *et al.*, 1976 : Jauncey, 1982 : De and Chatopadhyay, 2004) but studies on muscular content of DNA and RNA in food deprived fishes are limited. The juveniles of the red fish, *Sciarnops ocellatus*, had demonstrated that the levels of RNA/DNA were sensitive to environmental factors such as temperature, salinity and feeding regimes (Chung *et al.*, 1988, 1991, 1993).

The objective of this work is to determine nucleic acid levels in white and red musculatures and to use RNA/DNA ratio as representatives for physiological condition of mosquito fish *Gambusia affinis* in laboratory conditions, with the same feeding and starvation.

Materials and Methods

Full grown matured experimental fish were obtained from district anti-malaria department, in Akola city, acclimated to a laboratory conditions, then placed in experimental aquaria of 40 liters. They were fed commercial food *ad libitum*, or starved during a period of 15 days. After the experimental time had passed, 6 samples of each treatment were sacrificed to extract samples from the epaxial white muscle, dorsal musculature and the red muscle under the lateral line. Tissues were frozen until the moment of analysis. Determinations of RNA and DNA were made taking 15 mg of both tissues and were processed by the fluorometric method (Canino and Calderone, 1995). Two-way analysis of variance was used to determine the effects of feeding and the kind of tissue on the RNA and DNA levels.

Results and Discussion

The set of organisms submitted to fasting survived 15 days of the treatment. Analysis of variance revealed that there were significant differences between RNA and DNA levels and RNA/DNA ratios in both muscles types and feeding conditions, as well as an interactive effect between the variables (Table 21.1).

Table 21.1: Two Ways Analysis of Variance for Comparison of RNA, DNA and RNA/DNA Ratio in Muscles of *Gambusia affinis* Fed and Starved During 15 Days

Values	*Variations*		
	Among Tissue	*Feeding Condition*	*Interaction*
DNA	10.78**	4.01**	1.20
RNA	7.61**	16.41***	17.44***
RNA/DNA	5.74**	16.77***	17.50***

*$p < 0.05$; ** $p < 0.01$; *** $p < 0.001$, Ns: $p > 0.05$.

DNA levels were different with respect to the tissue, being major on red and white muscle of the no feeding fish (124.86±5.66 μg g^{-1} wet mass and 122.61±7.66, respectively), compared to red and white muscles of the feeding fish (117.66±20.66 and 99.66±0.55 μg g^{-1} wet mass, respectively (Table 21.2).

Table 21.2: Average Values (±SD) of RNA and DNA in White and Red Muscles of *Gambusia affinis* Fed and Starved for 15 Days

Experiment Conditions	*Muscles Types*	*DNA μg g^{-1}*	*RNA μg g^{-1}*
Fasting	White muscles	122.61 ± 7.66	114.00 ± 3.01
Fasting	Red muscles	124.86 ± 5.66	108.02 ± 9.93
Feeding	White muscles	99.66 ± 0.55	141.62 ± 9.66
Feeding	Red muscles	117.66 ± 20.66	112.83 ± 5.66

Thus, the feeding condition appeared to have a major effect on the DNA values. The basic levels of RNA/DNA in the white and red tissues of *Gambusia affinis* under conditions of feeding show that the white muscle tissue has a higher rate of growth than the red tissue. It is probably related to the white tissue's capacity to grow and to provide metabolic substrates to other tissues, while red musculature seems to be more related to the energetic contributions to fast movement of the fish. It is marked by a great variation over the energetic substrates of both tissues, and their mobilization is related to physiological processes such as instantaneous growth (Lemus *et al.*, 1993).

The levels of RNA/DNA ratio on the white musculature of the feeding organisms was 1.41±0.10, higher than that observed in the same tissue in conditions of initiation (0.48±0.17), demonstrating that there exists a significant increase on the rate of instantaneous growth. However, the red musculature showed less difference with high variation on the levels of RNA / DNA under feeding conditions.

References

Arthinton, A.H., 1989. Diet of *Gambusia affinis holbrookki, Xiphophorus helleri, X. maculatus* and *Poecilia reticulata* (Pisces : Poecilidae) in streams of southeastern Queensland, Australia. *Journal of Asian Fisheries Society*, 2: 193–212.

Chung, K.S., M. Nirchio, G.J. Holt and C. Arnold, 1988. Acidos nucleicos en la musculatura blancay roja de juveniles del pez rojo *Sciaenops ocellatus*. *Boletín del Instituto Oceanográfico de Venezuela* 27: 123–127.

Chung, K.S., J.G. Holt and C. Arnold, 1991. Influencia de la privación de la dieta y realimentación sobre el crecimiento y la relación ARN / ADN en los juveniles del pez rojo *Sciaenops ocellatus*. *Acta Científica Venezolana*, 42: 355–344.

Chung, K.S., G.J. Holt and C. Arnold, 1993. Influence of salinity and food deprivation on growth and RNA–DNA ratio in red drum *Sciaenops ocellatus* (Pisces : Scianidae). *Revista de Biología Tropical*, 41: 187–195.

De, D. and N. R. Chatopadhyay, 2004. Effects of dietary protein level on nutrient retention and body composition of Striped Catfish *Pangasius sutchi* (Fowler). *Environ. Ecol.*, 22: 349–352.

Fernando, A., V.P.E. Phang and S.Y. Chan, 1991. Diets and feeding regimes of poeciliid fishes. *Journal of Asian Fisheries Society*, 4: 99–107.

Garling, D.L. Jr. and R.P. Wilson, 1976. Optimum dietary protein to energy ratio for channel catfish fingerlings, (*Ictalurus punctatus*). *J. Nutr.*, 106: 1368–1375.

Jauncey, K., 1982. The effects of varying dietary protein level on the growth, food conversion, protein utilization and body composition of juvenile tilapias (*Sarotherodon mossanbicus*). *Aquaculture*, 27: 43–54.

Lumus, M., K S. Chung and G. J. Holt, 1993. Efecto de la temperatura sobre el crecimiento de juveniles de *Petenia kraussii* (Pisces : Cichlidae): Relación ARN / ADN. *Revista de Biología Tropical*, 41: 45–48.

Mills, D., 1984. *Community Fishes. A Fishkeeper's Guide*. Salamander Book Limited, London. 117 pp.

Mills, D. and G. Vevers, 1986. *Guía práctica ilustrada de los peces de acuario*. Editorial Blume S.A. Barcelona, España. 208 pp.

Olivier, A. and H. Kaiser, 1997. A comparison of growth, survival rate, and number of marketable fish produced of swordtails, *Xiphophorus helleri* Heckel (Family: Poeciliidea), between two types of culture system. *Aquaculture Research*, 28: 215–221.

Chapter 22

Toxicity Effect of Zinc on the Rice Field Crab *Spiralothelphusa hydrodroma*

S. Jayakumar, M. Deecaraman and S. Karpagam

Introduction

Accumulation of industrial effluents in water bodies leads to heavy metal pollution and cause serious damage to the life systems. The pesticides and heavy metals cause the greatest threat to the health of Indian aquatic ecosystem (Prasad, 2002) and in terrestrial ecosystem (Patnaik and Senapathi, 1996). The crustaceans are used to monitor metal levels in the environment (Sanders *et al.*, 1999). The field crab *Spiralothelphusa hydrodroma* and *Barytelphusa guerini* are the major component of the paddy field and has an edible importance among local population. Zinc is an essential metal but when the concentration exceeds the threshold level becomes toxic (Vijayaraman *et al.*, 1999).

Gills function as the major route for the uptake of heavy metals as they are the most permeable regions of the body (Victor, 1994). In the present investigation, the effect of zinc on freshwater field crab *Spiralothelphusa hydrodroma* was studied extensively. The morphology and histology of gills, hepatopancreas, were studied since these organs are directly effected by the toxic components. The changes in biochemical content and bioaccumulation rate were also studied.

Materials and Methods

The freshwater field crab *Spiralothelphusa hydrodroma* were collected from paddy field of perambakkam, Thiruvellore district, Tamil Nadu. The crabs were collected from irrigation channels and agricultural land at early morning hours by handpicking and stored in plastic containers. Intermoult crabs in stage III of ovarian development were selected, acclimatized, fed with uncooked oats meal. Crabs with carapace length 3.01 to 4.52 were chosen and were kept in plastic tanks partially immersed.

The toxicity studies were carried out with $ZnSO_4$ and the LC_{50} values for 24, 48, 72 and 96 h were calculated by Abotts formula (1925). Sublethal studies were carried out to assess the response of the test organism to stress caused by heavy metals. According to Sprague (1971) 1/3rd and 1/10th of the 96 h LC_{50} value represents higher and lower sublethal concentration and the study period was for 15 and 30 d.

The experimental crabs were sacrificed at the end of study period and dissected and the tissues namely gills and hepatopancreas were fixed in Bouin's fluid, processed and embedded in paraffin wax. Sections of 5-10 μm were cut, stained in Harris haematoxylin and eosin (Bancroft and Cook, 1984), observed under light microscope and photomicrographed. Protein content was estimated by Bradford (1976) method. Lipid content was estimated by Folch *et al.* (1957). Bioaccumulation studies were done as per the method of Van Hoof and Van san (1981). Statistical analyses were done by using WINKS 4.65 software.

Results and Discussion

The control crabs were active and had normal feeding behaviour, whereas in treatment the crabs were restless. The LC_{50} value for 96 h was 243.73 ppm. Botton *et al.* (1998) observed that the horse shoe crab *Limulus poluphenus* were able to survive up to 190 mg/L of zinc. He also observed that the embryo and larvae were highly resistant to heavy metals than the marine crustacea.

The gills are of phyllobranchiate type that consists of pairs of flattened lamellae branching from central axis and is deep brown in colour. Each gill plate was made of a double layer of cuticle enclosing gill epithelium. Central axis contained haemocytes and fixed nephrocytes.

In zinc treated crabs, gills were damaged even in 15 d of treatment. Thickening of gill epithelium, necrosis, vacuolation, deformation deeply staining granules were observed as in Figure 22.1. At higher concentration, the damage was very severe, gill lamellae were highly necrosed, swollen and central axis was highly distorted and highly vacuolated. Gill rachis was enlarged with ulceration and hyperplasia. Reduced haemocytes, high vacuolation, large number of deeply stained granules were observed.

Damages in the gills were observed in *Varcinus maenas* exposed to copper (Nonnotte *et al.*, 1993; Lawson *et al.*, 1995). They observed extensive structural alterations namely cellular hyperplasia, vacuolation, necrosis, thickening of gill epithelium, reduction of haemal spaces and tissue hypoxia. Similar changes in *Macrobrachium. idea* in response to cadmium and mercury was noted by Victor *et al.* (1990). They observed oedematous, necrosed lamellae with extensive vacuolation, ulceration and hyperplasia.

Gate and Mulherkar (1979) reported infiltration of haemocytes in gills of prawn *M. kistensis* exposed to copper. Oedematous and necrosed gill lamellae lead to loss of osmotic barrier and hyperplasia that reduced the respiratory ability. Shrinkage of respiratory surface in the gills causes severe respiratory failure (Victor, 1994). Depletion in the rate of oxygen consumption as a result of heavy metal toxicity might be the reason for mortality (Bansal, 1979; Rao *et al.*, 1998).

In control crabs hepatopancreas were dark yellow in colour. Histological studies revealed the presence of large number of elongated tubules united together by a connective tissue. Three types of epithelial cells namely R, B and F cells were observed. In treated the hepatopancreas were reduced in size, in collapsible condition and pale yellow in colour, tubules formed loose bunches, connective tissue was damaged. In 15 d treatment, the lumen size increased and vacuolation was more tubules were in distorted condition as in Figure 22.2. In 30 d of treatment, the damages were more pronounced, lumen size increased, connective layer was damaged, epithelial cells were reduced in length. Tubules were distorted and vacuolation was more. Size of nuclei was also reduced. The secretory cells and haemal space had increased in size but there was reduction in the haemocytes. Cell distortion and more vacuolation were seen.

Victor *et al.* (1990) observed histopathological changes in the hepatopancreas of *P. hydrodromous* in response to cythion resulting in reduction in the height of tubular epithelium, enlargement of lumen, vacuolation and atrophy. Similar disorganized condition of hepatopancreas was observed in *Uca. annulipes* in response to cadmium and mercury toxicity (Suresh, 2001). Extensive vacuolation of hepatopancreas was observed in *M. malcolmsonii* in response to endosulphan (Bhavan and Geraldine, 2000).

Zinc treated crabs showed reduction in protein content at both the lower and higher sublethal concentration. Decrease was statistically significant. Similarly reduction in lipid content was also observed (Table 22.1).

Table 22.1: Protein and Lipid Content of Zinc Treated Crabs

Organ	*Treatment Period*	*Control*	*Lower Sublethal Concentration*	*Higher Sublethal Concentration*	*F. Value*	*P. Value*
			Protein			
Hepatopancreas	15 days	71.16±1.11	65.41±1.72	63.42±2.49	32.57	<0.001*
	30 days	71.39±0.91	63.30±1.84	59.32±1.28	104.58	<0.001*
Gills	15 days	58.49±1.35	53.23±1.63	48.86±1.42	74.00	<0.001*
	30 days	58.66±1.05	50.12±1.52	47.86±1.57	86.6	<0.001*
			Lipid			
Hepatopancreas	15 days	91.08±1.12	68.26±1.09	57.01±1.22	1397.30	<0.001*
	30 days	91.30±1.10	59.50±0.75	49.50±1.23	2583.74	<0.001*
Gills	15 days	09.78±1.18	08.62±0.57	07.36±0.69	15.88	<0.001*
	30 days	10.15±0.72	07.68±0.68	06.50±0.43	42.44	<0.001*

Mean±SD of six individual observations; Values are expressed mg/g wet tissue.

*Statistically significant (By one-way analysis of variance).

The decrease in protein content was drastic and was maximum for 30 d of treatment. Similar decrease in protein level was observed in *M. lamarrei lamarrei* in response to copper (Krishnamoorthy and Subramanian, 1995) and in *Barytelphusa guerini* exposed to zinc sulphate (Sarojini *et al.*, 1990) and chromium (Reddy and Venugopal, 1991). Saravanabhavan and Geraldine (1997) reported decrease in protein level in *M. malcolmsonii* when exposed to endosulphan. The depletion of tissue protein was due to diversification of energy to meet the impending energy demand under toxic stress and altered enzyme activities (Reddy, 1987).

The decline in tissue protein content suggested intensive proteolysis in the tissue, which contributes to the amino acids to be fed in to TCA cycle. Reduction in protein level and depletion of RNA was due to RNAase activity and was responsible for the depletion of RNA and protein (Shah and Dubale, 1983). Saxena *et al.* (1989) suggested the decline in protein level was due to decreased availability of energy required for protein synthesis. The decreased protein level is also due to reduced food uptake and growth.

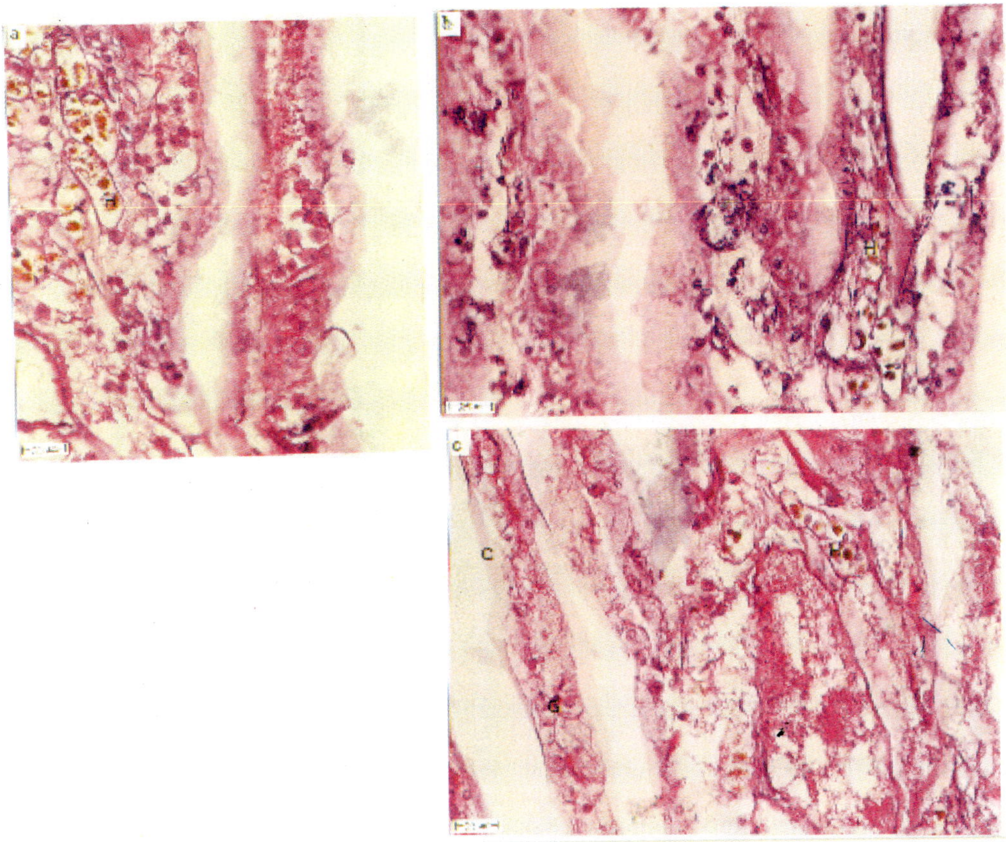

Figure 22.1: Gills: (*a*) Control; (*b*) 15d Treated; (*c*) 30d Treated
C: Cuticle; H: Haemocytes

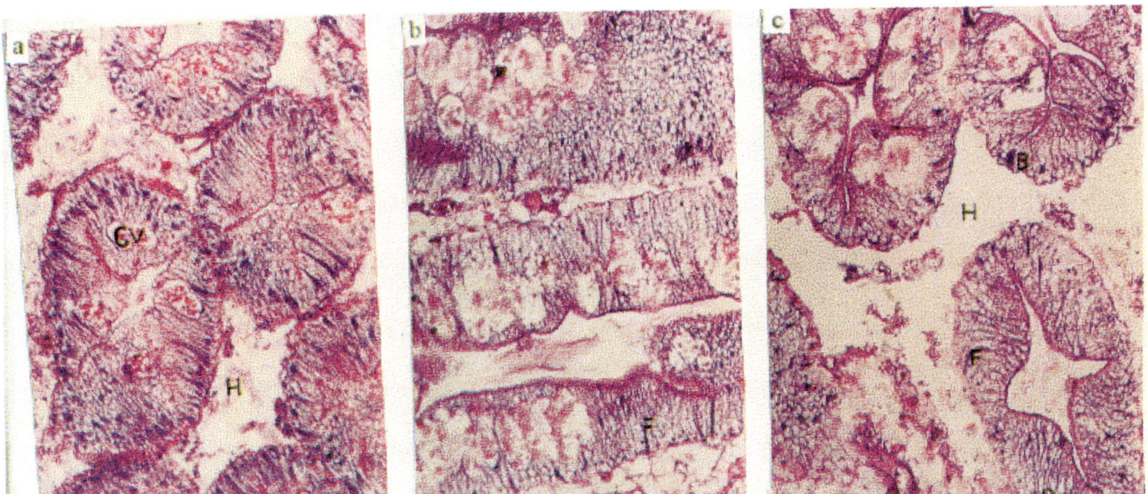

Figure 22.2: Hepatopancreas: (*a*) Control; (*b*) 15d Treated; (c) 30d Treated
Cv: Central vacuole; H: Haemal space

Nagabushanam *et al.* (1972) reported reduction in lipid content in hepatopancreas in *Metapenaeus idella* in response to pesticides. Reduction in lipid content was also reported in *Uca annulipes* exposed to cadmium and mercury (Suresh, 2001), *M. malcolmsonii* exposed to endosulphan (Saravanabhavan and Geraldine, 1997). The accelerated hydrolysis of lipid might be to cope up with the increased energy demand occurring due to metal toxicity. Among the biomolecules, protein and lipids represent the principal ones to be utilized at times of stress including metal toxicity for the derivation of energy. Their role in activation of energy in tissues has been elucidated in several studies on both invertebrates and vertebrates (Geraldine *et al.*, 1999).

Table 22.2: Bioaccumulation of Zinc in Hepatopancreas and Gills of the Crabs

Organ	*Treatment Period*	*Control*	*Lower Sublethal Concentration*	*Higher Sublethal Concentration*	*F. Value*	*P. Value*
Hepatopancreas	15 days	906.37±3.89	1749.0±4.39	4938.0±0.98	818387.6	<0.001*
	30 days	1012.97±3.47	2226.24±2.96	5172.38±5.16	173.83	<0.001*
Gills	15 days	762.22±7.37	1810.97±6.95	2994.26±6.91	149040.7	<0.001*
	30 days	772.15±5.98	2183.0±3.63	4210.08±4.21	803935.4	<0.001*

Mean±SD of six individual observations; Values are expressed μg/g wet tissue.

*Statistically significant (By one-way analysis of variance).

Bioaccumulation of zinc was greater in hepatopancreas and in gills also. The accumulation rate was time dependent. Though gills function as the major route for the uptake of heavy metals as they are the most permeable regions in the body, hepatopancreas acts as the major site of accumulation, metabolic activities and detoxifying function (Gibson and Baker, 1979). Essential heavy metals were stored as phosphate granules in hepatopancreas so that the concentration in the haemolymph was kept normal. Bryan (1986) suggested that the role of hepatopancreas was to curtail the entry of excess amount of heavy metals in to the haemolymph. The crustacean hepatopancreas is a vital organ for diverse metabolic activities such as synthesis, secretion, uptake of nutrients, excretion, metabolism and storage of inorganic reserves.

Regulation of excess metal concentration in tissues was accomplished through several means such as temporary absorption, storage and release by hepatopancreas; losses across the gills, through antennal gland, urine, faeces and in the form of metal binding proteins such as metallothionein. Zinc was known to be the constituent of carbonic anhydrase, an enzyme found in the blood, the muscle utilizes this enzyme more than the other tissues and there in consumes a greater quantum of zinc. Depuration may reduce zinc level but after extended period of exposure leads to biomagnification to higher levels (Vijayaraman *et al.*, 1999).

Aquatic invertebrates naturally accumulate abnormally high amounts of heavy metals. The effects of these heavy metals on the normal function of cells, tissues and organs are deleterious due to cumulative toxicity. Zinc is hazardous when accumulated beyond the trace level even though it is the essential metal found in the system of all living organism. Results of the present study show that the essential metal zinc cause damages to the tissues at higher sublethal concentrations.

References

Abott, W.S., 1925. A method of computing the effectiveness of an insecticide. *J. Econ. Entomo.*, 18: 265–267.

Bancroft, J.D. and Cook, H.C., 1984. *Manual of Histology Techniques*, (Ed.) Livingstone, C. London.

Bansal, S.K., 1979. Pesticide induced alterations in the oxygen uptake rate of a freshwater major carp *Labeo rohita*. *Ecotoxicol. Environ. Saf.*, 3: 374.

Bhavan, P.S. and Geraldine, P., 2000. Histopathology of the hepatopancreas and gills of the prawn *Macrobrachium malcolmsonii* exposed to endosulfan. *Aquat. Toxicol.*, 50: 331–339.

Botton, M.L., Johnson, K and Helleby, L., 1998. Effects of copper and zinc on embryos and larvae of the horse shoe crab, *Limulus polyphenus*. *Arch. Environ. Contam. Toxicol.*, 35(1): 25–32.

Bradford, M.M., 1976. A rapid sensitive method for the quantification of microgram quantities of protein utilizing principle of protein dye binding. *Anal. Biochem.*, 72: 248–254.

Bryan, G.W., Hummerstone, L.G. and Ward, E., 1986. Zinc regulation in the lobster *Homarus gammarus*. Importance of different pathways of absorption and excretion. *J. Mar. Bio. Ass. U.K.*, 66: 175–199.

Folch, J., Lee, S.M. and Solane-Stanely, G.H., 1957.A simple method for isolation and purification of total lipids from animal tissues. *J. Biol. Chem.*, 226: 497–508.

Gate, H.V. and Mulherkar, L., 1979. Histological changes in the gills of two freshwater prawn species exposed to copper sulphate. *Indian J. Exptl. Biol.*, 17: 838–840.

Geraldine, P., Saravanabhavan, P., Kaliamurthy, T. and Zayappragassarzan, Z., 1999. Effects of dichlorvos intoxication in the freshwater prawn, *Macrobrachium malcolmsonii*. *J. Environ. Biol.*, 20: 141–148.

Gibson, R. and Baker, P.I., 1979.The decapod hepatopancreas. *Oceanogr. Mar. Biol. Ann. Rev.*, 17: 285–346.

Krishnamurthy, P and Subramanian, P., 1995. Biochemical variation during accumulation and depuration of copper in *Macrobrachium lamarrei lamarrei* (H.M.Edwards). *Bull. Pure App. Sci.*, 14(1): 27–33.

Lawson, S.L., Jones, M.B. and Moate, R.M., 1995. Effect of copper on the ultrastructure of the gill epithelium of *Carcinus maenas* (Decapoda : Brachyura). In: *Proceedings "Trace Metals in the Aquatic Environment"*. Proc. of the III International conference, (Eds.) Aarthus. Philips, D.J.H. and Rainbow, P.S., 31(1–3): 63–72.

Nagabhushanam, R. and Deshpande, J. and Sarojini, R., 1972. Effect of some pesticides on the biochemical constituents of freshwater prawn *Macrobrachium kistensis*. In: *Proc. Nat. Symp. Ecotoxicol.*, p. 73–84.

Nonnotte, L., Biotel, F. and Truchot, J.P., 1993. Waterborne copper causes gill damage and haemolymph hypoxia in the shore crab *Carcinus maenas*. *Can. J. Zool. Rev. Can. Zool.*, 71(8): 1569–1576.

Patnaik, A. and Senapathi, B.K., 1996. Impact of Copper fungicide on the survivability, growth and generation of an Anecic earthworm. *Ecol. Env. and Cons.*, 2: 109–113.

Prasad, B.B., 2002. Effect of copper and Zinc in the gills of *Channa marulius*. *J. Ecotoxicol. Environ. Monit.*, 12(1): 35–40.

Rao, L.M., Vani, S. and Ramaneswari, 1998. Metal accumulation in tissues of *Macrobrachium rudae* from Mehadrigedda stream, Visakhapatnam. *Poll. Res.*, 17(2): 137–140.

Reddy, P.M., 1987. Toxic impact of malathion on the branchial protein metabolism of freshwater fish, *Cyprinus carpio*. *Environ. Ecol.*, 5: 368–369.

Reddy, S.L.N. and Venugopal, N.B.R.K., 1991. *In vivo* effects of cadmium chloride on certain aspects of protein metabolism in tissues of a freshwater field crab *Barytelphusa guerini*. *Bull. Environ. Contam. Toxicol.*, 46(4): 583–590.

Sanders, M.J., Du-Preez, H.H. and Van-Vuren, J.H.J., 1999. Monitoring cadmium and zinc contamination in freshwater systems with the use of the freshwater river crab, *Potamonautes warreni*. *Water S.A.*, 25: 91–98.

Saravanabhavan, P. and Geraldine, R., 1997. Alterations in concentrations of protein, carbohydrates, glycogen, free sugar and lipid in prawn *Macrobrachium malcolmsonii* exposed to sublethal concentration of endosulfan. *Pesticide Biochem. Physiol.*, 58: 89–101.

Sarojini, R., P.R. Machale and R. Nagabhushanam. 1990. Adverse effect of zinc sulphate exposure to ovarian development of freshwater crab, *Barytelphusa guerini*. *J. Nat. Conserv.*, 2: 97–104.

Saxena, P.K., Singh, V.P., Kondal, J.K. and Soni, G.L., 1989. Effects of some pesticides on *in vitro* lipid and protein synthesis by the liver of the freshwater teleost, *Channa punctatus* (Bl.). *Environ. Pollut.*, 58: 273–280.

Shah, P.H. and Dubale, M.S., 1983. Biochemical changes induced by malathion in the body organs of *Channa punctatus*. *J. Anim. Morphol. Physiol.*, 30: 107–118.

Sprague, J.B. 1970. Measurement of pollutant toxicity to fish. III: Sublethal effects and safe concentrations. *Wat. Res.*, 5: 245–266.

Suresh, V. 2001. A. study on the effects of heavy metals toxicity on a brackish water crab, *Uca* (*celuca*) *lacteal annulipes* (Crane, 1975) of pulicat lake, Tamil Nadu. Ph.*D. Thesis*, University of Madras, Tamil Nadu, India.

Van Hoof, F. and Van San, M., 1981. Analysis of copper, zinc, cadmium and chromium in fish tissues. A tool for detecting metal caused fish kills. *Chemosphere*, 10(10): 1127–1135.

Victor, B., 1994. Gill tissue pathogenecity and haemocyte behaviour in the crab *Paratelphusa hydrodromous* exposed to lead chloride. *J. Environ. Sci. Health*, 29A: 1011–1034.

Victor, B. Narayanan, M. and Nelson, D.J., 1990. Gill pathology and haemocyte response in mercury exposed *Macrobrachium idea* (Heller). *J. Environ. Biol.*, 11(1): 61–65.

Vijayaraman, K., John, G., Sivakumar, P. and Mohammed, R.R., 1999. Uptake and loss of heavy metals by the freshwater prawn, *Macrobrachium malcolmsonii*. (Milne Edwards.). *Indian J. Exp. Biol.*, 36: 1120–1124.

Chapter 23

Impact of Zinc and Cadmium on Hepatopancreas and Gonad of Estuarine Oyster *Crassostrea cattuckensis*

G.D. Suryawanshi

Introduction

The hepatopancreas is a main organ of storage of metabolic reserves, which provides a source of energy during the periods of physiological stress in addition to its role in the digestion of food. The impairment of the cytoarchitecture of hepatopancreas due to heavy metal stress definitely reflects in its metabolic disorder. This view is strongly supported by the work Bayne *et al.* (1987) observed the feeding and digestion by *Mytilus edulis*. They provided a food, mixture of slit and algal cell at low concentrations and found out the rate of feeding, the passage of time for food in the gut, absorption efficiency and metabolic rates. Mandal and Ghose (1970) studied the effect of calcium arsenate on the histopathology of giant African snail, *Achatina fulica*. Mix and Sparks (1971) on the repair of digestive tubules of *Crassostrea gigas* damaged by ionizing radiation. Manna and Ghose (1972) have studied the histopathological changes in the gut of the same animal caused by endrin. Morrigomez *et al.* (1990) reported histopathological changes in the digestive gland–gonad complex of the marine prosobranch, *Littorina littorina* exposed to cadmium. While Krishankumar *et al.* (1990) studied the cellular responses to copper and mercury in the digestive cells, cilia and digestive tubules of *Perna viridis*. Muley (1990) studied the fluoride-induced changes in the gonad and hepatopancreas of freshwater mussel *Indonea caeruleus.* Kulkarni (1993) reported effect of cadmium metal in the gonad and hepatopancreas of freshwater mussel *Lamellibranch marginalis*. Bielefeld (1991) histologically observed the digestive gland of starved bivalve from two different depths (2 and 9 m respectively). The activity of the digestive gland was stronger till ten days of starvation about 25 per cent of the gland tissue was degenerated and regeneration of gland was stronger after five days of refeeding. Kenneth (1993) has described the functional morphology of the digestive system of Lyonsia hylina. According to him this bivalve has a type stomach, which contains major typhosole that does not penetrate the left pouch. Patil and Mane (2000) studied the changes in hepatopancreas of the bivalve *Lamellidens marginalis* due to acute mercury stress.

On the other hand, heavy metals interfere with production of new generation in a variety of ways. Destruction of spawning grounds is one way in which shellfishes and many invertebrates are harmed. The cell damage in organisms may be due to some diseases or due to the effect of toxic substances. Such damage often undergo unnoticed before a significant damage to a internal organs is already caused. A study on pathological changes in the microanatomy of tissues is called as histopathology and it can give a useful data concerning tissue changes prior to external manifestation. Thus, examination of tissues of bivalve molluscs could give an early indication of pollution and this may help in avoiding future disaster. The use of histological technique to investigate pollution effects is hence in increasing use. Heavy metal poisoning or toxicity results in damage of normal structure at cellular level. Since heavy metals are known to interfere with functional groups of macromolecules, the presence of heavy metals above a threshold level results in irrevocable alteration in the microenvironment of the cell, thus resulting in its morphological changes towards degeneration. Therefore, a few authors have been performed experiments on representative aquatic animals both marine and freshwater for such studies (Singh and Sahai, 1984; Sastry and Sunita, 1984).

The reproduction in molluscs of most economic value is of special interest due to their importance as food for human. Although knowledge of reproductive cycle is necessary for predicting annual recruitment, interpreting, growth mortality and survival data and also in mariculture of the species. In bivalve molluscs many workers have studied the reproduction and much of the work has been reviewed exclusively on marine species and much of the literature on reproduction in bivalve molluscs is concerned with the gonad development and reports on breeding periods (Sastry 1979). Several environmental factors have been shown to control the reproductive cycle. Gametogenesis begins shortly after growth and maturation of the gonad. Maturation of the gametes is under several environmental factors such as temperature lunar periodicity, depth, mechanical factors, light intensity and a few others and also endogenous such as genetics and hormonal controls (Mackie, 1984). The maturation of gonad is also dependent on the richness of food supplies, which in turn are dependent on climate and other environmental factors. Amongst these exogenous factors some act as environmental clue and synchronizer in regulating reproductive activities (Giese and Pearse, 1974). For marine species the studies comprise by dividing the gonad into a number of stages based on microscopic examination of histological sections in different parts of the year. On *Donax cuneatus* Rao (1967), on *Donax faba* Algarswamy, (1969), on *Crassostrea gryphoides* Mane and Nagabhushnam (1976) and the subject has been reviewed for mussels and oysters by Mane (1997).

The present study was undertaken to understand the changes in the hepatopancreas and gonad of oyster *Crassostrea cattuckensis* under zinc and cadmium stress in different time exposure period and its detoxification mechanism.

Materials and Methods

Ratnagiri is southern district of Maharashtra state bearing productive estuaries and creeks. Bhatye estuary is one of them (situated 73°5 East and 16°51 North) in South of Ratnagiri coast where Kajvi river meets the sea. As the river Kajvi brings in a large load of mud, silt and suspended matter the water becomes turbid in monsoon, however the deposition of excessive sedimentation produces mud flats. These mud flats support rich vegetation and mangrove plants grown along the banks of river and near the river mouth the mangrove island exists. Since past eight to twelve years much deposition of the suspended matter and mud has been occurring over the submerged rocks in the estuary. In fact these rocks form the preferred substrata for oysters crassostrea to settle and grow. The oyster's settlement substrata are now depleted and only a few areas in the estuary are left for the oysters to grow. Since the preferred substrata in this estuary are totally submerged at a depth of about three to seven meters, they never get exposed to atmosphere during the tidal cycles. It is thought that rich mangrove areas in the estuary Bhatye are most favorable for oyster growth since they grow as much as 130 mm in shell length, the large size not common generally in other areas of the coast of Ratnagiri district. Along the banks of the river, far interior in the estuary and along the estuary the human population residing is mostly of the fishermen community and the mechanized boats for fishing in the sea pass regularly through this estuary. Among the bivalve molluscs the oysters, *Crassostrea cattuckensis* and *Saccostrea cucullata* along with others like *Katelysia opima*, *Meretrix meretrix* and *Sanguinolaria diphos* are found abundantly in the estuary and they provide delicious meat and valuable lime to the local people. Except during peak monsoon (July-August) the oysters *C. cattuckensis* are regularly fished and marketed.

During study the oysters *C. cattuckensis* were fished from Bhatye estuary in post winter (February). Soon after the fishing they were brought to the laboratory and the shells were brushed to clean the fouling biomass and mud. They were then stocked in continuous aerated filtered seawater pumped in the laboratory from the estuary for 24 to 48 h for depuration. Based on the acute toxicity tests the chronic test were performed on the oysters of (80-90mm shell length) using sub lethal concentrations, *i.e.* $1/10^{th}$of LC_0 and LC_{50} values of zinc and cadmium metal salts (*i.e.*0.65 and 0.94-ppm for zinc and 0.10 and 0.38-ppm for cadmium respectively). The experiments were carried for 15 and 30 days to metal exposure and then transfer of oysters of 15 days metal exposed to normal seawater for next 15 days. The control group in normal seawater was run simultaneously during experiment. The oysters belonging to control, and experimental were sacrificed separately to obtain hepatopancreas and gonad and these were fixed in Bouin's Hollande fixative for 48 hours. After fixation the tissues were then washed in distilled water and dehydrated in ethyl alcohol in different grades, cleared in xylene, embedded in tissue mat (at 58-60°c melting point) and these were then sectioned at 5-6 μm thickness on rotary microtome. These sections were stained with Mallory triple stain and mounted in DPX. All the observations and microphotography were made under the Labo VT-20 Videoscan research microscope.

Results

The changes in hepatic tubules and its cellular structure due to metal toxicity and detoxification of oysters along with control are shown in Figure 23.1. In normal (control) group of oysters the hepatopancreas consists of ducts and digestive tubules grouped in the form of bundles indistinctly separated and connected interlobular connective tissues of collagenous fibers. Each tubule is bounded by thin muscle fibres, which form the basement membrane. Each digestive tubule consists of digestive cells or columnar type, vacuolated and acidophilic and pyramidal type and basophilic type cells. In present study, when the oysters exposed to metal concentrations of zinc and cadmium the residual bodies or fragmentation spherules were comparatively more in the lumen in lower concentration than in higher concentration.

The results showed that the cadmium caused deteriorative changes in both concentrations comparatively zinc metal concentrations. During 15 days exposure period in higher concentrations of zinc and cadmium the tissue showed swelling of the tubules, which was distinct from the connective tissue. The nuclei of the digestive and secretory cells stained dark and most the cells detached from the basement membrane and from each other. The number of amoebocites in the connective tissue decreased considerably. However, during 30 days exposed period in lower concentration of both metals the tissue resulted the tubules completely lost their original shape probably due to dissolution of the basement membrane, which resulted in the dispersion of the digestive and secretory cells and formation of vacuoles. These cells mostly showed necrosis. The amoebocites increased in number in inter lobular spaces and many surrounded the tubule cells. On the other hand the oysters exposed to 15 days for metal accumulation those oysters was transferred to normal seawater for metal detoxification for next 15 days. It is pertinent to note that the detoxified oysters apparently regenerate the shape of new tubules after complete destruction of tubules during accumulation period in both metal concentrations.

The changes in gonad follicles and follicle streak due to metal toxicity and during detoxification process of oysters along with control group of oysters are shown in Figure 23.1. In normal (control) group of oysters showed many germ cells and follicle streak. Generally the oyster was spawned during study therefore the clear picture was not seen in control group and never seen the sperm morulae and spermatocytes and spermatids. The observations of the gonads showed the lumen was expanded in the metal accumulation period of 15 days as well as 30 days exposure period. Further the female ovaries many vitellogenic oocytes were found and being released in a few follicles degenerating oocytes were also seen in 15 days exposed oysters. During 30 days exposure period in lower and higher concentrations of zinc and cadmium metals the follicle wall ruptured at places and follicular content showed necrotic condition. This was more pronounced in LC_{50} group than the LC_0 group follicle remained compact. Where as during 15 days exposure period showed many previtellogenic and vitellogenic oocytes were formed. Prominent vacuole appeared in their nuclei and lipid globules and nutritive cells decreased much in quantity.

On the other hand the oysters exposed to 15 days for metal accumulation those oysters was transferred to normal seawater for metal detoxification for next 15 days. It is pertinent to note that the detoxified oysters contracted the expanded shape of lumen and apparently regenerate follicle streak of in both metal concentrations.

Discussion

Histopathology is an indispensable and powerful technique in establishing routine toxicology studies performed for the purpose of assessment of edible resources to human beings. Its value lies not only in the sensitivity in terms of toxic levels but particularly in the disclosure of large organs and mechanisms of action. To assess its usefulness in toxicology studies with fish and shellfish, several investigators have performed experiments using various environmental contaminants (Wester, 1991). It is in this view that the present study was done on the histological changes of depot tissue, hepatopancreas and reproductive organ gonad of the oyster *C. cattuckensis* exposed to zinc and cadmium and detoxified oysters under laboratory conditions. The experiments were conducted in post winter season. In *C. cattuckensis* the hepatopancreas consists of blindly ending tubules, which are linked with the stomach by a systems of branched, particularly ciliated ducts. The tubule epithelial contains two types of cells recognizable in the light microscope, one acidophilic, columnar and vacuolated and the other pyramidal and basophilic. The acidophilic cells are responsible for intracellular digestion of food and are usually referred to as digestive cells. These two types are also recognized in the present study. The literature on the digestive gland of bivalve molluscs is extensive and Yennawar (1997) studied the changes in the organ with the tide and described. A review of Owen (1972) is particularly useful because it illustrates the structure of the tubule cells of *M. edulis* and discusses the role of digestive cells in terms of the formation and breakdown of lysosomes, a concept previously applied to the bivalve digestive tubules by Sumner (1969). Lysosomes are membrane bound vesicle containing

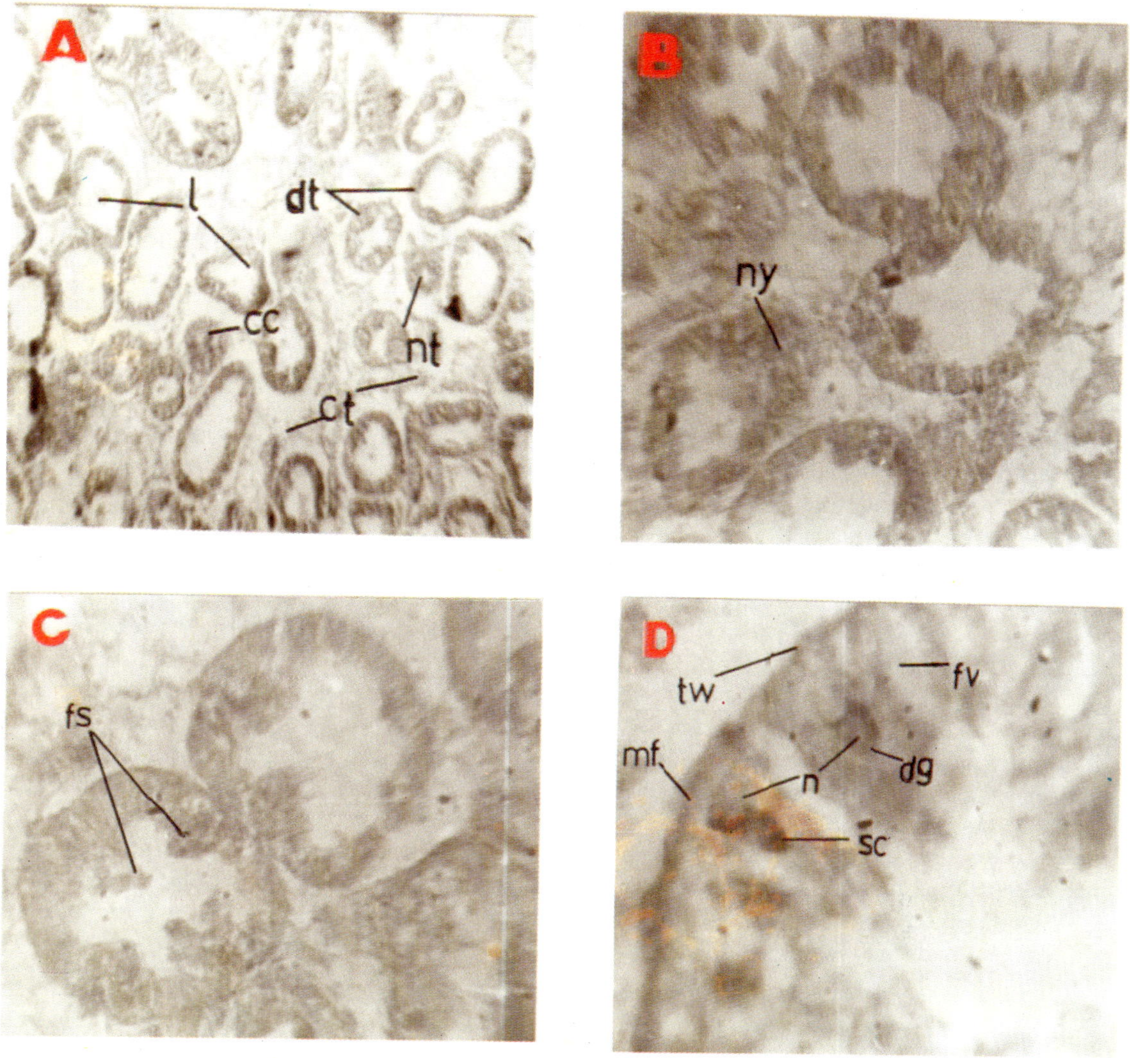

Figure 23.1: Hepatopancrease Control, A-100x, B-200x, C-400x, D-1000x

L: Lumen; Ct: Connective tissues; Dt: Digestive tubules; Fv: Food vacuoles, Dg: Digestive gland; Sc: Secretory cells; n: Nucleus; tw: Tubular wall; fs: Fragmentation spherules; cc: Clumps of cells; nt: New tubules; ny: Nest of young cells; mf: Muscle fibre.

hydrolytic enzymes. These vesicles serve as intracellular disposal sites for metals. An important general cytotoxic response to contaminants is a change in the stability of lysosomal membranes. This change results in lysosomal hydrolase's being released to the cytoplasm and can be induced at toxicant concentrations lower than those that cause acute effects. Thompson *et al* (1974) stated that the hepatopancreas or digestive gland serves as site for the storage of metabolic reserves and provides a source of energy during the periods of physiological stress in addition to its role in digestion of food. The destruction of cellular structure of hepatopancreas due to pollutants definitely reflects in the metabolic activity of it. However, in *M. edulis* no significant reduction in lysosomal integrity was evident in the group exposed to the lowest copper concentration, however, significant reduction in lysosomal latency was observed in the group exposed to highest concentration of copper as observed by Harrison Berger (1982). Akarte (1985) observed the pesticides like folithion and lebaycid were effective in all the seasons to *Indonea caeruleus* and *L. marginalis* to cause cellular destruction of hepatic lobules in different seasons. The changes were loss of connective tissue, shrinked or desquamated basement membrane, loss of orientation of digestive or secretory cells, vacuolization or swelling of cells and in more pronounced in season.

In the present study on the oyster *C. cattuckensis* vacuolization and increase in intracellular space within the tubule was observed than in the control group. The digestive and secretory cells were both affected at cellular levels at different intensities the destruction was more. Pronounced in higher concentration of zinc and cadmium than lower concentration and more affected during after 30 days exposure than the 15 days exposure period, which might be due to higher intensity of metals, effect in higher concentration and long duration period. Further in detoxification process produced profound changes. The regenerative changes of cells were observed in the bivalves when it was exposed to lower and higher concentrations for 15 days those oysters transferred in normal sea water for next 15 days. Kenneth *et al.* (1986) they have found cadmium associated with all sub cellular fractions and the cytosol. Cytosolic cadmium was associated with high molecular weight metallothionein ligand pools but not the very low MW ligand. They also observed for total cadmium accumulation, cadmium in the sub cellular fractions and cytosolic pools. Through out the entire range of exposures there were no shifts in cadmium accumulation or its relative distribution among the sub cellular fractions in polychaetes. Moore (1991) stated that pathological reactions of the lysosomal system in hepatopancreatic cells of bivalve molluscs have proven to be sensitive bioindicators of pollutant effect. The author studied the effects of several cell-to-cell signaling compounds on the lysosomes of the hepatopancreatic digestive cells. The results showed the presence of cell surface receptors for epinephrine, acetylcholine and prostaglandin–F2- alpha, indicating that regulation of digestive cell lysosomes is unlikely to pose a problem for the use of lysosomal reactions as biomarkers for chemical insult.

Whereas in present study, the destruction of the basement membrane or tubule wall and cell surface of the digestive cells after metal exposure probably caused dysfunction of the surface receptors as quoted by Moore (1991) resulting in the disturbance of the lysosomal system functioning. Bigas *et al.* (1997) studied the oyster's display a diversity of uptake mechanisms for metallic elements and distribution in the target organs, namely gills and the digestive gland. During experimental exposure all animals survived the treatment. Mercury accumulation in gill tissue with a maximum concentration after 25 days of exposure, mercury levels were lower in remaining tissues, in which the maximum concentration was reached after 18 days of exposure, after these times, concentration in both tissues decreased. Results show that oysters can accumulate mercury from the environment, without their survival being affected during the experimental period. Structural alterations of epithelial tissues of gill and digestive gland of other metallic elements in bivalve molluscs. Interstial tissue was disorganized in the digestive gland and ultra structural changes in intracellular endomembranes were detected in the epithelial cells of the digestive gland after 30 days of treatment. Absorptive epithelial cells of gills showed highly dilated, swollen microvilli. These intracellular alterations are parameters of the incipient response to accumulation of mercury. Marigomez *et al.* (1990) carried out histological investigation of the digestive gland–gonad complexes of winkles that were experimentally exposed to sub lethal concentrations of cadmium. The structural integrity of digestive and basophilic cells of the digestive gland drastically changed such that both cell type become morphologically indistinguishable at the light microscopic level. The volume densities of haemocytes and pore cells located both ground the stomach and in interstitial connective tissues were significantly increased due to cadmium exposure. In the present study the drastic changes observed in the hepatopancreatic tubules probably due to the metals. Domouthtsidou and Dimitriadis (2000) have studied the intracellular localization of heavy metals using utomettalography. In the gills localized in the dense bodies observed in apical and in the basal part of the cells. Mercury presented the highest accumulation, followed by the Ag, Pb and Cu. The heavy metal exposure also affected the gross morphology of the examined tissues and resulted in the fusion of residual bodies in the digestive cells, the fragmentation or vacuolization of the rough endoplasmic reticulum and the increase in the number of granules in the basophilic cells. In the gills, fusion of the gill filament was also observed.

A response potentially related to lysosomes is the accumulation or inorganic and organic contaminants in membrane bound vesicles. These vesicle appear to be formed from the fusion of endocytosed metals with primary lysosomes. When metals are sequestered and immobilized in these vesicles, they may be detoxified. The present study was undertaken the metal detoxification by hepatopancreas and those results showed accumulated metals within 15 days, it was significantly detoxified within next 15 days.

The effect of contaminants on genetic material is of special interest because alterations in chromosomes in somatic tissue may be an important factor in the induction of neoplastic tumors. Furthermore, alteration in chromosomes in germ cells may result in adverse effect on reproductive success from gamete death or from the induction of dominant. One of the important aspects of these general responses to stress is the effect on pollution dynamics and community composition. Changes in reproductive potential are particularly relevant at these higher levels of biological organization. Early development can also be used as a measure of general stress because it represents a very sensitive stage of the life cycle. It is a period of rapid growth and cell division and particularly sensitive to limitations of energy resources. Moreover, it encompasses a number of events, including fertilization, setting and organogenesis that are particularly sensitive to stressors. Perturbations in any of these processes can result in arrested or abnormal development. Cajarville *et al.* (1991) while studying on the mussel *M. galloprovincialis* exposed to crude oil and commercial lubricant oil indicated that lysosomes showing cytochemically demonstrable *b*- glucoronidase activity were smaller and much more numerous in oocytes of mussels treated with 40 per cent dose of crude oil and lubricant oil when compared to controls. The author suggested that the structure of the lysosomal systems of oocytes was different from that of somatic cells (i. e. digestive cells). These changes in the lysosomal compartment appear to be associated to the process of gametes release or spawning.

The metal compounds attack the phospolipid structures of the plasma membrane and cause structural damage to them Valle and Ulmer (1982). Further, these compounds are known to be extraordinarily lipophytic, they bind to the sulphahydril group of many proteins Verma *et al.* (1983). The active site of ATPase contains the–SH bond to which mercury readily binds the enzyme becomes denatured Sahana *et al.* (1986). As a result of these properties mercury enter the cells, diffuses into the cytoplasm, attacks other cellular organelles especially mitochondria and lysosomes and causes inactivation of several vital enzyme systems. The structural damage caused by mercury to the lysosomal membrane may cause rupture of lysosomes releasing all hydrolytic enzymes leading to the cell autolysis Hossain and Dutta (1986). Earnshaw *et al.* (1986) studied the action of heavy metals on the gametes of *M. edulis*. Treatment of the mussel sperms with external concentrations of copper or zinc caused a decreased in motility in which zinc was more inhibitory than copper. Zinc appeared to cause more extensive mitochondrial damage than did treatment with copper. In the present study, it is possible that penetration of heavy metals in the gonad tissue of *C. cattuckensis* could have resulted in the changes in lysosomal compartment of vitellogenic oocytes, particularly in higher concentration and 30 days exposed oysters due to which the mature gametes instead of being spawned out showed degenerative changes. In addition, the genotoxic effect of cadmium on the gametes due to metals in *C. cattuckensis* also cannot be ruled out. It is important to note the data obtained by Dube and Dnforesme (1990) who found that in *M. edulis* showed oocytes are arrested at metaphase I of meosis until fertilization. This metaphase arrest requires the continuous synthesis of short-lived protein, the destruction of which is sufficient to induce meosis resumption followed by DNA synthesis. Based on these findings it can be presumed under present study that vitellogenic oocytes formed due to metal stress are retained on metaphase Ist and no short-lived proteins are synthesized because of the formation of metalloproteins. In addition, it is possible that the total or partial destruction on the synthesis of vitellin- like protein from the haemolymph and related ovarian proteins could have caused the little large sized oocytes to undergo lysis after metal stress. It has been reported that the achivement of gametogenesis in adult bivalve molluscs, *M. edulis* during reproductive cycle depends upon the action of different neuroendocrine factors, a mitotic factor allowing the multiplication of oogonia and spermatogonia, a meiotic factor triggering the first meotic division in males, a previtellogenic factor leading oocytes to vitellogenesis and a vitellogenic factor acting on vitellogenesis Mathiew and Lubet (1980).

Further in the present study abnormalities developed in the development of gonadal cells, leading to the appearance of enhanced growth of these cells of *C. cattuckensis* could be also due to the underlying mechanisms in the functional changes brought about by the different duration of exposure period. The oysters exposed to 15 days for metal accumulation those oysters was transferred to normal seawater for metal detoxification for next 15 days. It is pertinent to note that the detoxified oysters showed shape of new tubules, reform the digestive cells after complete destruction of tubules during accumulation period in both metal concentrations apparently regenerate the digestive gland during detoxification process. Whereas the gonad contracted the expanded shape of lumen and swelling of previtellogenic oocytes were less comparatively in accumulation period apparently regenerate follicle streak of in both metal concentrations. Organisms have evolved various strategies that appear to lower the effect of environmental changes, these include avoidance responses, repair or stabilization mechanisms and synthesis of detoxification enzymes or binding ligands. Presumely these strategies have evolved with a minimal cost in energy or rate limiting nutrients, although all of these processes require energy or nutrients that are diverted from other uses. The present study was under taken the gonadal changes in cellular and sub cellular levels of 15 days metal accumulated oysters and those oyster results showed drastic changes observed in next 15 days detoxified oysters.

This study focused on understanding how oysters adapt to and are affected by the wide range of concentrations of essential and non-essential metals in the aquatic environment. The coastal marine environment, where concentrations of essential trace metals vary from limiting to toxic levels, particularly challenges an organism's ability to maintain appropriate reservoirs of essential metals. This problem is often compounded by the presence of significant concentrations of non-essential metals that can compete with essential metals for uptake sites and intracellular binding sites. Such competition reduces the concentration of essential metals that are effectively available to the organism. To deal with these challenges, marine organisms have evolved a number of adaptive strategies that are collectively referred to as metal metabolism. These strategies include the active regulation of metal uptake, distribution, metabolism and excretion.

Acknowledgements

The author is grateful to Prof. U. H. Mane, Director, Marine research Laboratory, Ratnagiri for providing necessary facilities and also thankful to ICAR, New Delhi for the award of fellowship.

References

Akarte, 1985. Effect of organophosphorus insecticide on bivalve molluscs. *Ph.D. Thesis,* Marathwada University, Aurangabad, M.S., India, p. 1–190

Bayne, B.L., Hawkins, A.J.S. and Navarro, E.L., 1987. Feeding and digestion by the mussel *Mytilus edulis* (L.) (Bivalve Mollusca) in mixture of silt and algal cells at low concentrations. *J. of Exptal. Mar. Biol. and Ecol.,* 111: 1–22.

Bigas, M., Amiard-Triquet, C., Durfort, M. and Poquet, M., 1997. Sublethal effects of experimental exposure to mercury in European flat oyster *Ostrea edulis*:Cell alterations and quantitative analysis of metal. *Biometals.,* 10(4): 277–284.

Bielefeld, D.V., 1991. Histological observation of gonads and digestive gland in starving *Dreissena polymorpha* : *Bivalvia. Malacologia,* 33: 31–42.

Cajarville, M.P., Dtez, G.M., Marrigomez, T.A. and Angulo, E., 1991.Responses of basophilic cells of the digestive gland of mussels to petroleum hydrocarbon exposure. *Dis. Aqut. Org.,* 9(3): 221–228.

Domouthtsidou, G.P. and Dimitriadis, V.K., 2000. Ultastructural localization of heavy metals (Hg, Ag, Pb, and Cu.) in gills and digestive gland of mussels, *Mytilus galloprovincialis. Arch. Environ. Contam. Toxicol.,* 38(4): 472–478.

Dube, F. and Dnforesme, L., 1990. Release of metaphase arrest by partial inhibition of protein synthesis in blue mussel oocytes. *J. Expt. Zool.,* 256(6): 322–332.

Earnsaaw, M.J., Wilson, S., Akberali, H.B., Bulter, R.D. and Marriott, K.R.M., 1986.The action of heavy metals on the gametes of the marine mussel, *Mytilus edulis* III the effect of applied copper and zinc on sperm mortality in relation to ultrastructural damage and intracellular metal localization. *Mar. Environ. Res.,* 20(4): 261–278.

Giese, A.C. and Pearse, J.S., 1974. Introduction. In: *Reproduction of Marine Invertebrates,* (Eds.) Giese A.C. and Pearse, J.S. Academic Press, New York, p. 2–93.

Hussain, A. and Dutta, H.M., 1986. Acid phosphatase activity in the intestine and the caeca of blue gill exposed to methyl mercury chloride. *Bull. Environ. Contam. Toxicol.,* 36: 460–467.

Kenneth, A.T., 1993. The functional morphology of the digestive system of *Lyonsia hyaline* Conrad, 1831 (Bivalvia, Anomalodesmatas, Pandorroidea). *J. Molluscan Stud.,* 59: 175–186.

Kenneth, D., Jenkins and Brenda, M. Sanders, 1986.Relationships between free cadmium ion activity in sea water, cadmium accumulation and subcellular distribution and growth in polychaetes. *Environ. Health. Perspectives.,* 65,205–210.

Kulkarni, S.D., 1993. Cadmium toxicity to freshwater bivalve Molluscs *Lamellidens marginalis* from Godavari river near Aurangabad. *Ph.D. Thesis,* Marathwada University, Aurangabad.1–338.

Mackie, G.L., 1984. Bivalves. In: *The Mollusca Vol. 1*, (Ed.) Wilbur K.M. Academic Press, New York, London, p. 351–402.

Mandal, T.K. and Ghose, K.C., 1970. Application of calcium Arsenate Histopathological changes and glycogen mobilization in *Achatina fulica*.Bowdich. *Indian J. Exp. Biol.*, 8(4): 332–333.

Mane, U.H. and Nagabhushnam, R., 1976. A study on the reproductive Biology of Indian oyster *C. gryphoides. Marathwada Univ. J. Mar. Sci.*, 5: 243–246.

Mane, U.H., 1997. Mussels and Oysters. In: *Fouling Organisms of theIndian Ocean: Biology and Control Technology*, (Eds.) R. Nagabhushnam and Thompson. Oxford and IBH publishing Co. Pvt. Ltd., New Delhi, Calcutta, p. 441–506.

Manna, B. and Ghose, K.C., 1972. Histopathological changes in the gut of *Achatina fulica* Bowdich caused by Endrin. *A molluscicidi. Indian J. Exp. Biol.*, 10(6): 461–463.

Marrigomez, J.A., Cajaraville, M.P. and Angulo, E., 1990. Histopathology of the digestive gland and gonad complex of the marine prosobranche *Littorina littorina* exposed to cadmium. *Aquatic. Org.*, 9(3): 229–238.

Mix, M.C. and Spark, A.K., 1971. Repair of digestive tubule tissue of the pacific oyster, *Crassostrea gigas* damaged by ionization radiations. *J. Invert. Pathol.*, 17: 172–177.

Moore, M.N., 1991. Lysosomal changes in the responses of molluscan hepatopancreatic cells, exticellular signals. *Histochem. J.*, 23(10): 495–500.

Muley, D.V., 1990. Fluoride induced changes in the gonad and hipatopancreas of freshwater mussel, *Indonaia caeruleus.* Absracts *11th Ann. Sess. Acad. Environ. Biol.*, Aurangabad.

Owen, G., 1972. Lysosomes, peroxisomes and bivalve. *Science Progress*, 60: 299–318.

Patil, S.S. and Mane, U.H., 2000. Changes in hepatopancreas of the bivalve *Lamellidens marginalis* due to acute mercury stress. *Proc. Nat. Acad. Sci. India*, 70(B), II: 133–138.

Rao, K.R., 1967. Annual reproductive cycle of the wedge clam *Donax cuniatus* (L). *J. Mar. Biol. Assoc. India*, 9: 141–146.

Sahana, S.S., Jana, S., Choudhary, M.A. and Choudhary, D.K., 1986. Hg (II) induced changes in some biochemical parameters in the freshwater fish *Clariuas batracus. Physiologica, Bohemslovaca*, 35: 81–85.

Sastry, A.N., 1979. Pelecypoda. In : *Reproduction of Marine Invertebrates*, (Eds.) Giese, A.C. and Pearse, J.S. academic press, New York, San Francisco, 5: 123–265.

Sastry, K.V. and Sunita, S., 1984. Chronic effect of chromium in *Channa punctatus. J. Environ. Biol.*, 5(1): 47–52.

Singh, S. and Sahai, S., 1984. Effect of malathion on the mortality and behaviour of two freshwater teleost. *J. Environ. Biol.*, 5(1): 23–28.

Sumner, A.T., 1969. The distribution of some hydrolytic enzymes in the cells of the digestive gland of certain Lamellibranches and gastropods. *J. Zool.*, 158: 277–291.

Thompson, R.J., Ratcliffe, N.A. and Bayne, B.L., 1974.Effect of starvation on structure and function in the digestive gland of the mussel, *Mytilus edulis. J. Mar. Biol. Assoc.*, U.K., 54: 699–712.

Valle, B.L. and Ulmer, D.D., 1982. Biochemical effects of mercury, cadmium and lead. *Ann. Rev. Biochem.*, 41: 91–128.

Verma, S.R., Madhur Jain and Tonk, I.P., 1983. *In vivo* effect of mercuric chloride on tissue ATPase of *Notopterus notopterus*. *Toxicol. Lett.*, 16: 305–310.

Wester, P.W., 1991. Histopathological effects of environmental pollutants B. HCH and methyl mercury on reproductive organs in freshwater fish. *Comp. Biochem. Physiol.*, 110 C(1/2): 237–239.

Yennawar, P.L., 1997. Some aspects on reproductive biology of west coast bivalve molluscs. *Ph.D. Thesis*, Dr. Babasaheb Ambedkar Marathwada University, Aurangabad (M.S.), p. 1–394.

Chapter 24

The Pathobiology of Intestinal Neoechinorhynchosis, Caused by *Neoechinorhynchus rutili* in Freshwater Fish

G. Benarjee and B. Laxma Reddy

Introduction

The acanthocephalan members are endoparasitic and occur in the digestive tract of terrestrial and aquatic vertebrates, ranging from fish to man, but occurs quite mostly in fish (Schaperclaus, 1991). There are several instances of extremely heavy acanthocephalan infection in fish (Bullock, 1962 and 1963, Aitzetmuller *et al.*, 1994, Weber *et al.*, 1994 and 1995). The earliest workers who have studied on the acanthocephala parasites are Boyd (1951), Bullock (1963) and Hine and Kennedy (1974). Kennedy and Rampus (1976) reported the distribution of acanthocephalan, *Pomphorhynchus laevis* along the alimentary tract of several species of fresh water fishes. Kennedy and Lord (1982) while studying the acanthocephalans from the alimentary canal of fishes, observed the habitual specificity of *Acanthocephalus clavulus*, in the eel fish, *Anguilla anguilla*. Researchers who have contributed in this direction from other parts of the world are Samuel and Bullock, 1981; Amin, 1985 a, b; Wanstall *et al.*, 1986; Taraschewski *et al.*, 1987; Taraschewski, 1988, 1989a, 1989b; Taraschewski, 1989, Amin and Heckman 1992; Amin and Nahhas 1994; Weber *et al.*, 1994; Aitzlemuller *et al.*, 1994; Weber, *et al.*, 1995; and Polzer and Taraschewski, 1993, 1994. But relatively much work has not been done on the Acanthocephalan members found in the fish fauna of Indian subcontinent. However the contributions made by Hilda Leiching, 1984; Ibragimov, 1989; Khan Aly and Shaukat, 1991 and Moravec and Scholz, 1991 are significant. Therefore a study has been carried out in this direction with an aim to visualize the impact of the parasitism of Acanthocephalan, *Neoechinorhynchus rutili,* on fresh water fish, *Clarias batrachus.*

Materials and Methods

For the present investigation fishes were procured from the local fish market and also collected directly from the fishermen who directly caught fish from the local fresh water bodies. To collect the parasites the fishes were sacrificed and screened after opening the alimentary canal. Since most of the parasites infect various organs of the alimentary canal, the entire alimentary tract was isolated from the fish and kept in petridish containing normal saline. Intestine was thoroughly screened for the presence of the parasites. The parasites were then collected on a slide and tied for fixation. The parasites were stained with Alumcarmine. The intestine of the infected and uninfected fish were isolated and preserved in Bouins, Susa, Carnoy and Zenkers fluids (Gurr, 1962) for the histopathological and histochemical studies (Pearse, 1968; Bancroft, 1975). A battery of histochemical tests were applied on the microtome cut sections of intestine both infected and uninfected to demonstrate and also to assess the histochemical changes that occur in the tissues of the infected organ if any due to infection with *Neoechinorhynchus rutili.*

Results

Neoechinorhynchus rutili is a common intestinal parasite of *Clarias batrachus* especially in the tropical region. The short spiny protrusible cylindrical proboscis of the parasite may be highly useful for the parasite to adhere. The histoarchitecture of the transverse section of intestine of the fish, *Clarias batrachus* reveals that it consists four typical layers, from outermost to the inner, Serosa, muscularis, submucosa and mucosa layer (Figure 24.1). The association of parasite with the host results the development of pathological state in the host intestine which is very much conspicuous. The pathological state starts with the attachment of parasite to the inner walls of the intestine (Figure 24.2). The pathogenecity of the worm is usually presumed on the basis of the depth of penetration of the proboscis or the intensity of pathogenecity may also presumed by number of worms found in the host.

The histopathology of the intestine particularly revealed the drastic histoarchitectural changes, this include severe damage to the villi and other layers of the intestine (Figure 24.3). In the present study it is noticed that the parasite after invaded the mucous membrane of the intestine has caused extensive damage to its villi. This resulted the eruption of villi, inflammation, fibrosis associated with hyperplasia and metaplasia (Figure 24.4). The parasite caused inflammation and dilatation of blood vessels in the submucosa (Figure 24.5). The chronic histopathological changes that are brought by the parasite in the host intestine includes vacuolation of submucous cells and necrosis of epithelial cells (Figure 24.6). However, the parasite could not be penetrated to the extent of muscularis layer. Thus this layer of intestine was intact. The most pronounced pathological changes that occur to the fish intestine was the thickening of intestinal muscularis layer. However because of the longer proboscis of the parasite the intestinal damage was more severe. The proboscis embeds deeply in the intestinal wall and it firmly attached to it thus it penetrated into the submucous layer. It was observed in the present study that the appearance of yellowish white granules and nodules in the outer intestinal wall. These are the connective tissue capsules developing around the proboscis while the parasite piercing the intestinal wall, it is presumed. The proliferations can be very extensive and can also damage the neighbouring tissues of the liver etc. The parasite therefore produced series of abnormalities in the intestine of the fish may be collectively considered as intestinal Neoechinorhynchosis. These pathological changes are however predominantly present at the posterior part of intestine than the anterior part could be supported that the frequent detachment separation and renewed firm attachment.

Table 24.1: Histochemical Tests for Intestine

Histochemical Tests Applied	*Results*	
	Un-infected	*Infected*
Periodic Acid/Schiff (PAS)	+	+++
PAS/Saliva	+	+++
Schiff's without oxidation	+	+++
Acetylation/PAS	-	-
Deacetylation/PAS	+	+++
Alcian blue 1.0 Ph	++	++
Alcian blue 2.5 Ph	++	++
Alcian blue 1.0 pH/PAS	++	++
Alcian blue 2.5 pH/PAS	++	++
Alcian blue/safranin	++	++
Alcian blue/Aldehyde fuchsin	++	++
Mercuric Bromophenol blue	+	+++
Ninhydrin/Schiff	+	++
Ferric ferricyanide	+	++
Congored	+	+++
p-DMAB nitrite	+	++
$KMnO_4$/Alcian Blue	++	+++
Millon's reaction	+	++
Copperpthalocyanin	++	+++
Sudan Black 'B'	+	+++
Orcein/Van Gieson	+	+++
Orcein	+	+++

+++: Strongly positive; ++: Moderately positive; +: Positive; –: Negative.

The battery of histochemical tests conducted on the sections of intestine of *Clarias batrachus* revealed an interesting picture on the histochemical nature of the tissues which may change the physiology of the organs. In the present study the histochemical studies conducted on intestine of infected fish gave a unique different picture when compared with the uninfected one. Periodic acid/Schiffs (PAS technique) yielded an intense positively in all the layers of intestine of the uninfected fish, which suggested the occurrence of Vinyl hydroxyl group in the tissues. Further this reaction also suggests the occurrence of mucopolysaccharides and other related carbohydrate groups. The distribution of carbohydrates in the intestine of infected fish with acanthocephalan, *Neoechinorhynchus rutili* revealed a different picture on the presence of food substances at the intestine part of the digestive tract. The changes in the distribution of carbohydrates therefore shown relationship between the host and the parasite (Figure 24.7). Even the glycogen content of the intestine showed a difference from the uninfected fish, this was evidenced when the section were subjected to PAS after diastase digestion. In the infected fish the total carbohydrate and glycogen content are increased significantly. This increase of carbohydrates may be the protective interaction developed by the fish to combat with the parasitic effect. While the sections of intestine of uninfected fish showed a positive response to Alcian blue both at 1.0 pH and 2.5 pH in the infected fish reveals the presence of more quantity of mucopolysaccharides (Figure 24.8). This increase may be attributed to increase of mucous secretion in the intestine against the invade of the parasite. The mucous secretion is a protective interaction of the host against parasite. The strong positive reaction for Bromophenol blue gave a strong positivity indicating the presence of proteins. The intensity of the reaction is more in the infected fish compared with the uninfected fish. In the infected fish an increase in the various groups of lipid including total lipids and phospholipids. The histochemical changes that are noticed in the intestine after the application of histochemical tests were presented in the Table 24.1.

Discussion

In the present investigation the fishes were very often seen infected with *Neoechinorhynchus rutili* which is an indication suggests that the frequency and incidence of parasitic infection is more at this region. The occurrence of Acanthocephalan in the fish was studied (Samuel and Bullock, 1981; Amin,

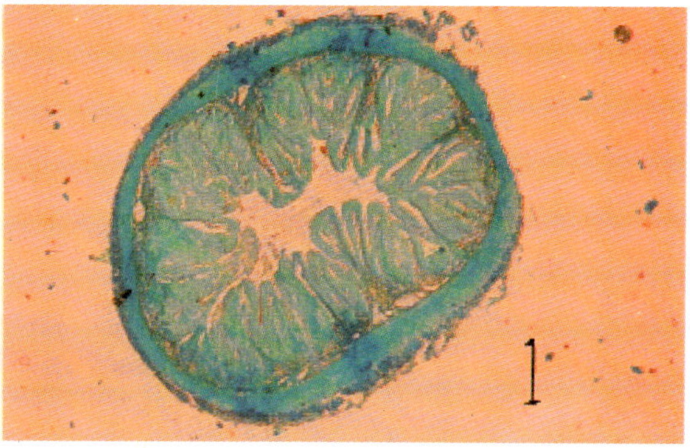

Figure 24.1: T.S. of Intestine of *Clarias batrachus*

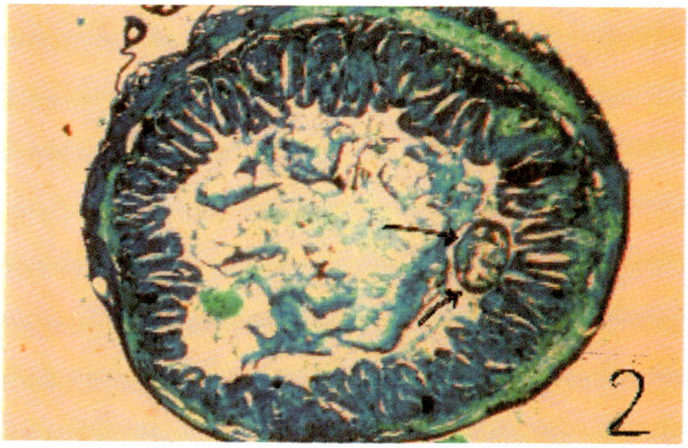

Figure 24.2: T.S. of Intestine of Infected Fish Showing Establishment of Association of the Parasite

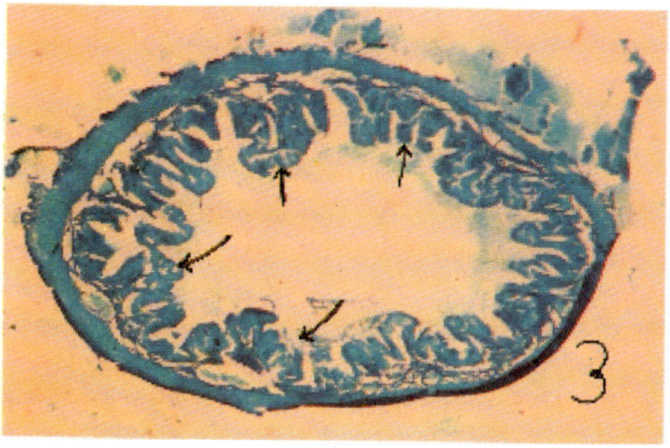

Figure 24.3: T.S. of Intestine of Infected Fish Showing Damage of Villi

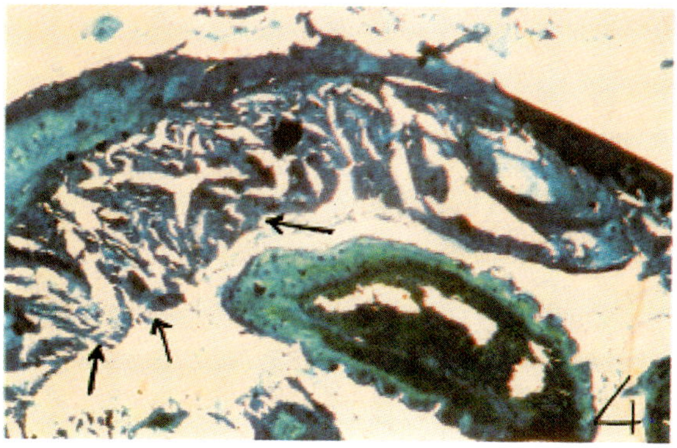

Figure 24.4: T.S. of Intestine of Infected Fish Showing Fibrosis Associated Hyperplasia and Metaplasia

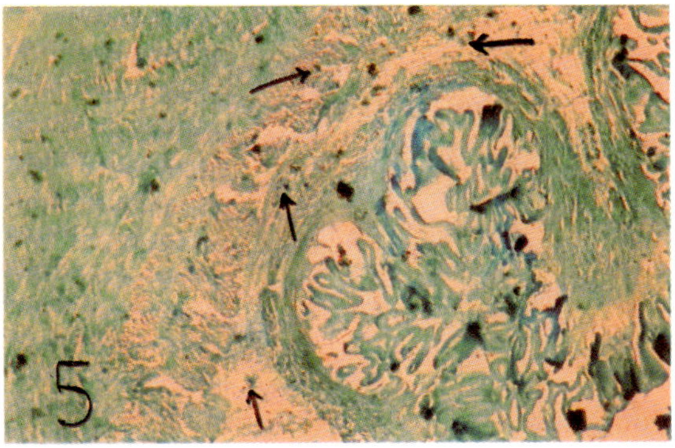

Figure 24.5: T.S. of Intestine of Infected Fish Showing Dilation of Blood Vessels and Damage Caused to the Submucous Membrane

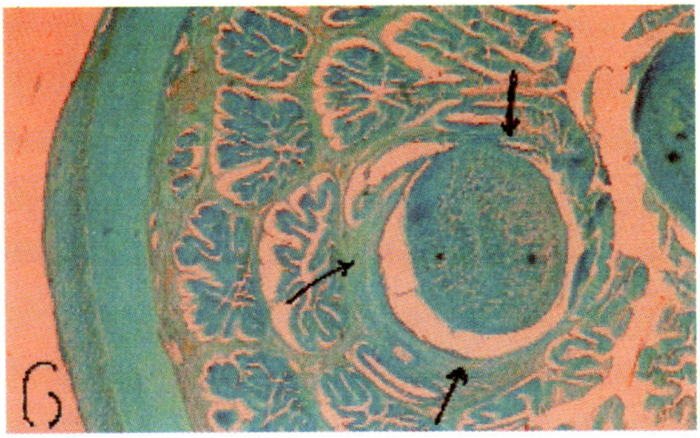

Figure 24.6: T.S. of Intestine of Infected Fish Showing Vacuolation of Submucous Membrane and Necrosis of Epithelial Cells

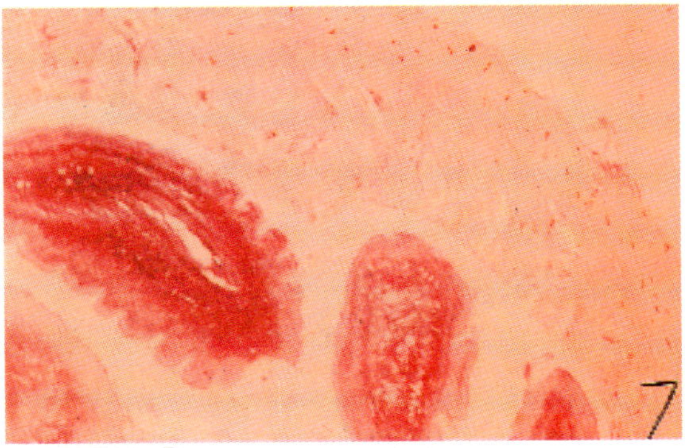

Figure 24.7: T.S. of Intestine of Infected Fish Stained with PAS

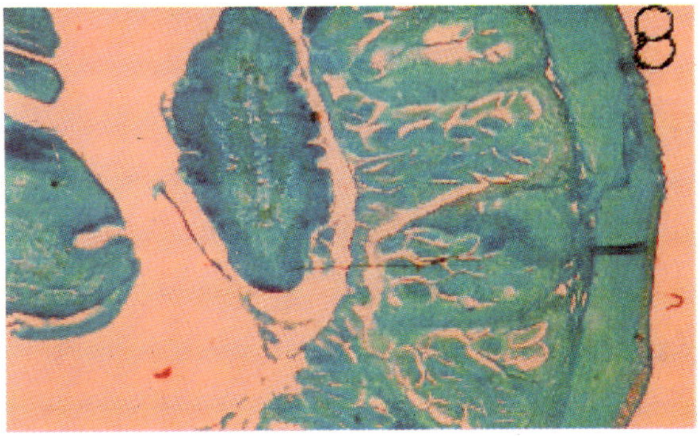

Figure 24.8: T.S. of Intestine of Infected Fish Stained with Alcian Blue 2.5 pH Showing the Presence of Mucopoly-saccharides

1985a, b; Taraschewski *et al.*, 1987; Taraschewski, 1989; Amin and Heckman *et al.*, 1994; Weber, 1994; and Amin and Nahhas 1994). This parasite exist in the intestine region of alimentary canal and they remain found in the same site throughout the period (Hoffmann, 1967). Acanthocephalans which are easily recognised by their typical hook-bearing proboscis are frequently encounter in the intestine of fishes. Adult acanthocephalan affecting fishes often cause extensive damage to the intestinal mucosa (Bullock, 1962 and 1963). The extent of damage caused to the intestine may be depended on the intensity of infection, which has been noticed in the present study. The parasite has brought histopathological alterations in the intestine of the host fish, due to which the abscess have been noticed in the intestine. The worms penetrated the intestinal wall through the mucosa, submucosa and muscularis layer. Due to this penetration inflammation of the intestine has been noticed, thus it produced histopathological alterations of the intestine. The pathogenecity of these spiny headed worms is usually presumed on the basis of the depth of penetration of the proboscis or the presence of large number of worms in the host. There are several instances of extremely heavy acanthocephalan infection in fish with much severe pathogenecity. Chaicharan and Bullock (1967) studied the histopathological changes in the *catostomid fishes* infected with acanthocephalans. Wanstall *et al.*, 1986; Polzer and Taraschewski, 1994 and Taraschewski, 1988, 1989a, 1989b have reported that the various species of acanthocephalan members found infected to rainbow trout revealed that they could produce histopathological abnormalities in the gut of the fish. The histopathological changes observed in the host organs may be either due to the mechanical damage or due to toxic secretions of the parasite. The changes that occurred in the intestine may be due to mechanical damage which led to erosion of the intestinal tissue. In contrary to this an increase of thickness in the muscularis layer may be considered as an adaptation of the host to combat the presence of the parasite. Further the destruction of other organs of the host suggests that there is a poor supply of nutrients by the infected digestive tract. This situation further leads to onset for deterioration of fish health. The degenerative changes in the infected organs are due to the secretion of proteolytics enzymes which may involved in the degradation of host tissues. The trypsin, a proteinase enzyme secreted by the acanthocephalans may digest the collagen found in the tissues and bring degenerative changes in the gastrointestinal tissues of the host. (Polzer and Taraschewski, 1994). The tissue damage caused to the intestine of *Clarias batrachus* may be due to the secretion of proteolytic enzymes by the acanthocephalan and caused degenerative changes in the intestine. Damage to the intestinal epithelium and proliferation of host connective tissue are the most pronounced histopathological effects. Wanstall *et al.* (1986) studied the pathological changes in rainbow trout induced by *Pomphorhynchus laevis*. Heavy acanthocephalan infection in fish with much vigorous pathogenecity was observed and it caused similar histopathological damage as found in *salmonoid* fishes (Bullock, 1963). However the pathogenecity of acanthocephalans in fish is not as simple as the other worms (Boyd, 1951; Bullock, 1962). The assessment of pathogenecity of a parasite can be made by the study of morphological and histochemical alterations of the tissues.

As a matter of fact when a parasite brings histopathological changes in the tissue may increase or decrease various metabolites. A similar state have been noticed in the present study. The acanthocephalan interfere with digestion, absorption, and food intake that causes metabolic disorders. Various secretions and metabolites of the parasite may cause toxic conditions in the intestine of fish. These abnormal conditions influences the secretory and digestive nature of the intestine. The histochemical nature of intestine reveals its physiological functional state and the pathological conditions certainly reflects on a change in the histochemical nature which certainly influences by and large the functional status of the organ. The parasite live in the intestine of the fishes also interfere with the protein metabolism of the host due to which the host tissue may show a decrease of increase in the level of the protein content. Therefore the quantity of protein depends on the metabolic activity of the body and cellular integrity. The strong positive reaction for Bromophenol blue gave a strong positivity indicating the presence of proteins. The intensity of the reaction is more in the infected fish compared with the uninfected fish. The increase in the protein content in the infected fish may be due to more supply of nutrients to the effected organs. However the quantum of increase depends upon the degree of infection, the degree of pathogenecity and the degree of resistance. Whereas the other proteins which were found positive to histochemical tests are Tyrosine, Tryptophan, glycoproteins, protein bound aminogroups and disulphides. The intensity of the reaction however varied which shows the degree of pathological condition prevailed in the fish due to parasitic infection. The lipid profiles of various organs and their distribution are influenced by the parasitic infection. Since the parasitic infection brings dietary change and metabolic process a change can be observed even in the various lipid contents of the organ. A similar condition has been noticed in the present study where in the infected fish an increased in the various groups of lipid including total lipids and phospholipids. The distribution of total lipid including total lipids and phospholipids was by copperpthalocyanin these staining reactions suggested the role of *Neoechinorhynchus rutili* in the rise of lipid content. The present histopathological state of the tissue clearly indicates that the worm might have impaired the routine physiological process of the fish.

Acknowledgements

The facilities provided by the Head, Department of Zoology is gratefully acknowledged. The authors express thanks to Prof. T.Rajendranath for critical evaluation of the manuscript and suggestions made for the necessary improvement of the discussion.

References

Aitzet Muller, K.H., Taraschewski, C. Filipponi and Weber, N., 1994. Lipids of fish parasites and their hosts: Fatty acids of phospholipids of *Paratenuisentis ambiguus* and its host eel (*Anguilla anguilla*). *Comparative Biochemistry and Physiology*, 109B: 383–389.

Amin, O.M. and Fuad M. Nahhas, 1994. Acanthocephala of marine fishes off Fiji Islands, with Descriptions of *filiosoma longcement glandatus* N. sp. (Cavisomidae) and Gravid females of *Rhadino rhynchus* johnstoni (Rhadinorhynchidae); and keys to species of the genera Filiosoma and Neorhadinorhynchus. *American Society of Parasitologists. J. Parasitol.*, 80(5): 764–774.

Amin, O.M., 1985a. Classification. In: *Biology of the Acanthcephala*, (Eds.) D.W.T. Crompton and B.B. Nickol. Cambridge University Press, London, p. 27–72.

Amin, O.M., 1985b. Acanthocephalan from lake fishes in Wisconsin. *Neoechinorhynchus robertbaueri* N.Sp. from Erimyzon sucetta (Lacepede) with a key to species of the genus *Neoechinorhynchus* Hamann, 1892, from North American fresh water fishes. *Journal of Parasitology*, 71: 312–318.

Amin, O.M. and Richard A. Heckmann, 1992. Description and pathology of *Neoechinorhynchus idahoensis* N.Sp. (Acanthocephala: Neo echinorhynchidae) in *Castomus columbianus* from Idaho. *J. Parasitol.*, 78(1): 34–39.

Bancroft, J.D., 1975. *Histochemical Techniques*. Butterworths, London and Boston.

Boyd, E.M., 1951. A survey of parasitism of the starting *Sturnus vulgaris* L. in North America. *J. Parasitol.*, 37: 56–84.

Bullock, W.L., 1962. A new species of Acanthcephalus from New England fish, with observations on variability. *J. Parasit.*, 48: 442–451.

Bullock, W.L., 1963. *Neoechinorhynchus rutili prolixoides* N. Sp. (Acanthocephala) from North American fishes. *Proceedings of the Helminthological Society of Washington*, 30: 92–96.

Bullock W.L., 1963. *Journal of Morph.*, 112: 23.

Bullock, W.I.., 1963. Intestinal histology of some salmonid fishes with particular reference to the histopathology of acanthocephalan infections. *J. Morph.*, 112: 23–44.

Chaicharan, A. and Bullock,W.L., 1967. The histopathology of Acanthocephalan infection in suckers with observations on the intestinal histology of two species of *catostomid fishes*. *Acta. Zool. Stockh.*, 481: 1–42.

Gurr, E., 1962. *Staining Animal Tissue: Practical Theoretical*. Leonard Hill (Books), Ltd. London, p. 631.

Hilda Leiching, 1984. Description of *Neoechinorhynchus salmonis* Sp.N. (Acanthocephala : *Neochinorhyn chidae*) from freshwater fishes of British Columbia. *J. Parasitol.*, 70(1): 286–91.

Hine, P.M. and Kennedy, C.R., 1974. Observations on the distribution, specificity and pathogenicity of the Acanthocephalan *Pomphorhynchus laevis* (Muller). *J. Fish. Biol.*, 6(4): 521–535.

Hoffman, G.L., 1967. *Parasites of North American Freshwater Fishes*. University of California Press, Berkele, California, 486 p.

Ibragimov, Sh.R., 1989. Thorny headed worms in caspian fish. IZV Akad. *Nauk. Az. SSR Ser. Biol. Nauk.*, 0(6): 66–69.

Kennedy, C.R. and Rampus, A., 1976. Long-term changes in the size of the *Pomphorhynchus Laevis* (Acanthocephala) population in the River Avan. *Journal of Fish Biology*, 10: 35–42.

Kennedy, C.R. and Lord, D., 1982. Habitat specificity of the acanthocephalan *Acanthocephalus clavula* is eels. *Anguilla anguilla. J. Helminthol.*, 56(2): 121–129.

Khan Aly, F.M. Bilquees and Shaukat, S.S., 1991. Seasonal variation in the occurrence of *Pallisentis ophiocephali* and *Acanthocentis betwai* (Acanthocephala) in relation to their fish hosts. *An Gew Parasitol.*, 32(3): 165–171.

Moravec, F. and Scholz, T., 1991. Observations on the biology of *Pomphorhynchus laevis* (Zoega in Muller, 1776) (Acanthocephala) in the Rokytna River. Czech and Slovak Republic. *Helminthologia*, 28: 23–29.

Pearse, A.G.E., 1968. *Histochemistry: Theoretical and Applied*, 2nd Edition. Little Brown and Company, Boston, MSS.

Polzer, M. and Taraschewski, H., 1993. Identification and characterization of the proteolytic enzymes in the developmental stages of the eel pathogenic nematode *Anguillicola crassus*. *Parasitology Research*, 79: 24–27.

Polzer,•M. and Taraschewski, H., 1994. Protelytic enzymes of *Pomphrhynchus laevis* and in 3 other Acanthocephalan species. *J. Parasitol.*, 80(1): 45–49.

Samuel, G. and Bullock, W.L., 1981. Life cycle of *Paratensisentis ambiguus* (Van cleave). *Journal of Parasitology*, 75: 911–919.

Schaperclaus, W., 1991. *Fish Diseases*, Vol. 2 Oxonian Press Pvt. Ltd., New Delhi, p 837–849.

Taraschewski, H., 1989. Host parasite interface of *Paratenuisentis ambiguus* (Eoacanthocephala) in naturally infected eel and in laboratory infected stickle backs and Juvenile Carp and rainbow trout. *Journal of Parasitology*, 75: 911–919.

Taraschewski, H., Moravec, F., Lamah, T. and Andrews, 1987. Distribution and morphology of 2 helminths recently introduced into European eel populations. *Anguillicola Crassus* (Nematoda: Darcunculoidae) and *Paratenuisentis ambiguus* (Acanthcephala, Tenuisentidae). *Diseases of Aquatic Organisms*, 3: 167–176.

Taraschewski, H., 1988. Host parasite interface of fish acanthocephalans. 1. Acanthocephalus anguillae (Palaeacanthocephala) in naturally infected fishes. LM and TEM investigations. *Diseases of Aquatic Organisms*, 4: 109–119.

Taraschewski, H., 1989a. *Acanthocephalus anguillae* in intra and extra intestinal position in experimentally infected juveniles of gold fish and carp and in stickle backs. *Journal of Parasitology*, 75: 108–118.

Taraschewski, H., 1989b. Host-parasite interface of *Partenuisentis ambiguus* (EO–acanthocephala) in naturally infected eel and in laboratory infected sticklebacks and Juvenile Carp and rainbow trout. *Journal of Parasitology*, 75(6): 911–919.

Wanstall, S.T., Robotham, P.W.I. and Thomas, J.S., 1986. Pathological changes induced by *Pomphorhynchus laevis* Muller (*Acanthocephala*) in the gut of rainbow trout. *Salmogairdneri Richardson Zietschrift fur parasitenkunde*, 72: 105–114.

Weber, N., Vosmann, K., Aitzet Muller, K., Filipponi, C. and Traschewski, H., 1994. Sterol and fatty acid composition of neutral lipids of *Paratenuisentis ambiguus* and its host eel. *Lipids*, 29: 420–427.

Weber, Taraschewski and Kurt Aitzet Muller, 1995. An elongation of fatty acids in the fish parasite *Paratenuisentis ambiguus* (Acanthocephala). *J. Parasitol.*, 81(4): 501–504.

Index